Atlas of Physarum Computing

ATLAS OF Physarum Computing

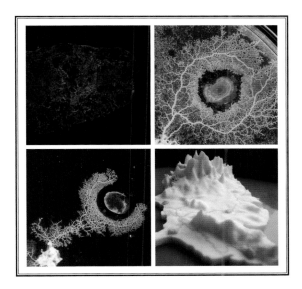

editor

Andrew Adamatzky
University of the West of England, UK

NEW JERSEY • LONDON • SINGAPORE • BEIJING • SHANGHAI • HONG KONG • TAIPEI • CHENNAI

Published by

World Scientific Publishing Co. Pte. Ltd.
5 Toh Tuck Link, Singapore 596224
USA office: 27 Warren Street, Suite 401-402, Hackensack, NJ 07601
UK office: 57 Shelton Street, Covent Garden, London WC2H 9HE

British Library Cataloguing-in-Publication Data
A catalogue record for this book is available from the British Library.

ATLAS OF PHYSARUM COMPUTING

Copyright © 2015 by World Scientific Publishing Co. Pte. Ltd.

All rights reserved. This book, or parts thereof, may not be reproduced in any form or by any means, electronic or mechanical, including photocopying, recording or any information storage and retrieval system now known or to be invented, without written permission from the publisher.

For photocopying of material in this volume, please pay a copying fee through the Copyright Clearance Center, Inc., 222 Rosewood Drive, Danvers, MA 01923, USA. In this case permission to photocopy is not required from the publisher.

ISBN 978-981-4675-31-4

Printed in Singapore

Preface

Unconventional computing is a journey to uncover principles of information processing and computation in physical, chemical and biological systems and to apply these principles in developing novel computing paradigms, architectures and implementations. This emerging field of science and engineering is predominantly occupied by theoretical research, e.g. quantum computation, membrane computing and dynamical systems computing. For a hundred theoretical papers there may be just one experimental laboratory paper. Only a handful of experimental prototypes are reported so far, for example, gas-discharge analogue path finders; maze-solving microfluidic circuits; geometrically constrained universal chemical computers; specialised and universal chemical reaction–diffusion processors; universal extended analogue computers; maze-solving chemotactic droplets; enzyme-based logical circuits; spatially extended crystallisation computers for optimisation and computational geometry; and molecular logical gates and circuits. A weak representation of laboratory experiments in the field of unconventional computers could be explained by technical difficulties, costs of prototyping of novel computing substrates and also psychological barriers. Chemists and biologists do not usually aspire to experiment with unconventional computers because such activity diverts them from mainstream research in their fields. Computer scientists and mathematicians would like to experiment but are "scared" of laboratory equipment. If there was a simple to maintain substrate, which requires minimal equipment to experiment with and whose behaviour is understandable and appealing to researchers from all fields of science, then progress in designing novel computing devices would be much more visible. There is such a substrate. This is the slime mould *Physarum polycephalum*.

Physarum polycephalum belongs to the species of order *Physarales*, subclass *Myxogastromycetidae*, class *Myxomycetes*, division *Myxostelida*. It is commonly known as a true, acellular or multi-headed slime mould. Plasmodium is a "vegetative" phase, a single cell with a myriad of diploid nuclei. The plasmodium is visible to the naked eye. The plasmodium looks like an amorphous yellowish mass with networks of protoplasmic tubes. It feeds on bacteria, spores and other microbial creatures and microparticles.

In our European project "Physarum chip: growing computers from slime mould", we designed and fabricated a distributed computing device built and operated by the slime mould *P. polycephalum*. In Physarum chips, data are represented by spatial configurations of attractants and repellents, or by geometrical constraints of the substrate, and the slime mould represents results of the computation with morphology of its protoplasmic networks or changes in electrical oscillator activity. Advantages of the Physarum chips are parallel inputs (optical-, chemo- and electrical-based) and outputs (electrical and optical) and their ability to solve a wide range of computational tasks, including optimisation on graphs, computational geometry, robot control, logic and arithmetical computing.

We fabricated a range of sophisticated machines made from the slime mould: sensors, image processors, logical gates, memristors, self-healing wires and memory devices. Most of these slime mould computers function in space and present results of their computation by changes in the morphological structure of their protoplasmic networks. This morphological computing behaviour is visually attractive and aesthetically appealing, and therefore we decided to share the best snapshots of slime mould computers with the World by preparing this atlas. In this unique set of photographs and micrographs, we illustrate in superb detail the nature of the slime mould computers and hybrid devices. Each entry includes a self-contained description of how the visualised phenomenon is used in the relevant slime mould computer, and has real scientific, theoretical or technological interest. This atlas is unique in providing the depth and breadth of knowledge in harnessing behaviour of the slime mould to perform computation. The atlas will help readers to understand how exploitation of biological processes has sparked new ideas and spurred progress in many fields of science and engineering.

Andrew Adamatzky
Bristol

Acknowledgements

Some works presented in the atlas were inspired or partially supported by the Leverhulme Trust research project "Mould intelligence: biological amorphous robots", the Samsung Global Research Outreach Program, the Next Generation Computing and Interconnections project "A feasibility study on interfacing nano- and macroworlds via amorphous biological computing substrate" and the European project FP7-ICT-2011-8 "Physarum chip: growing computers from slime mould". Richard Mayne thanks David Patton and David Corry (UWE, Bristol) for their technical expertise with the confocal and electron microscopes used, respectively. Jeff Jones acknowledges the support of the Department of Computer Science, UWE (Bristol) for PhD funding and the award of a SPUR early career researcher grant from UWE. Andrew Adamatzky thanks Soichiro Tsuda, of the University of Glasgow, for sending samples of slime mould's sclerotium in 2006. Andrew Adamatzky also thanks Dee Smart (UWE, Bristol) for curating his art exhibition "Slime mould machines" (@Bristol) in 2014; a catalogue of the exhibition was published as a stand-alone volume "The Silence of Slime Mould" (Luniver Press, 2014). Tomohiro Shirakawa was partially supported by a Grant-in-Aid for JSPS fellows No. 19-8731, a Grant-in-Aid for Research Activity Start-up No. 22800092 and a Grant-in-Aid for Young Scientists (B) No. 23700187 from the Japan Society for the Promotion of Science. Martin Grube is grateful to Christian Westendorf, Kathrin Helmel and Sigrun Kraker for support in the laboratory. We thank Michael Jones for editing the text and all the staff of World Scientific for their invaluable support.

Contents

Preface v

Acknowledgements vii

1. Slime mould gates, roads and sensors 1
   Andrew Adamatzky

2. Multi-agent model of slime mould for computing and robotics 35
   Jeff Jones

3. Slime mould biotechnology 47
   Richard Mayne

4. Slime mould interactions with chemicals and materials 61
   Benjamin De Lacy Costello

5. Basic features of slime mould motility 75
   Tomohiro Shirakawa

6. Slime mould grown on polymer layers 91
   Alice Dimonte, Tatiana Berzina and Victor Erokhin

7. Diversity of slime mould circuits 99
   Martin Grube

8. Slime mould fluids and networks from an artist's point of view   105
   Theresa Schubert

Bibliography   111

Index   115

Chapter 1

Slime mould gates, roads and sensors

Andrew Adamatzky

Unconventional Computing Centre,
University of the West of England, Bristol, United Kingdom
andrew.adamatzky@uwe.ac.uk

The photographs present a wide range of problems solved by the slime mould *P. polycephalum*: imitation of human-made transport pathways,[1] realisation of Boolean logical gates,[2] fabrication of self-repairing routable biowires,[3] implementation of delay elements in computing circuits,[4] computational geometry,[5] sensors[6,7,8] and a would-be nervous system.[9]

When inoculated on a substrate with scattered sources of nutrients Physarum propagates towards the sources and spans them with a network of protoplasmic tubes. The structure of the network may vary between experiments; however, statistically, the most common planar graphs approximated are proximity graphs.[10] When the configuration of nutrients matches a configuration of major urban areas of a country, the slime mould approximates a human-made transport network, i.e. motorways and highways of the country.[1]

On a nutrient substrate Physarum expands as an omnidirectional wave, e.g. as a classical excitation wave in a two-dimensional excitable medium.[5] It shows a pronounced wave front, comprising a very dense network of protoplasmic tubes. There are several orders of tubes which are differentiable by their width. The density of the protoplasmic network decreases towards the inoculation site, the epicentre of the wave pattern. Morphological transitions of the slime mould's networks during expansion, colonisation and development bear a remarkable resemblance to the Cosmic Web.[11] Web-like spatial arrangements of galaxies and masses into elongated filaments of the Cosmic Web[12] are represented by wave-fragment-like active growing zones and colonies of Physarum. Morphologies of sheet-like walls and dense compact clusters[12] are typical for the slime mould growing on a nutrient agar.

Maze solving is a classical task of bionics, cybernetics and unconventional computing. A typical strategy for maze solving with a single device is to explore all possible passages, while marking visited parts, till the exit or a central chamber is found. Several attempts have been made to outperform Shannon's electronic mouse Theseus[13] using propagation of disturbances in unusual computing substrates, including excitable chemical systems, gas discharge and crystallisation. Most experimental prototypes were successful yet suffered from the computing substrate's specific drawbacks. One of the entries of the chapter illustrates our laboratory experiment on path finding with Physarum guided by diffusion of an attractant placed in the target site.[14]

Given a cross-junction of agar channels and plasmodium inoculated in one of the channels, the plasmodium propagates straight through the junction;[2] the speed of propagation may increase if sources of chemoattractants are present (however, the presence of nutrients does not affect the direction of propagation). An active zone, or a growing tip, of plasmodium propagates in the initially chosen direction, as if it has some kind of inertia. Based on this phenomenon we designed two ballistic Boolean gates with two inputs and two outputs.[2] In these gates input variables are x and y and outputs are p and q. Presence of a plasmodium in a given channel indicates TRUTH and absence indicates FALSE. Each gate implements a transformation $\langle x, y \rangle \to \langle p, q \rangle$.

A growing slime mould can develop conductive pathways, or wires, with its protoplasmic tubes.[3] Given two pins to be connected by a wire, we place a piece of slime mould on one pin and an attractant on the other pin. Physarum propagates towards the attractant and thus connects the pins with a protoplasmic tube. A protoplasmic tube is conductive, can survive substantial over-voltage and can be used to transfer electrical current to lighting and actuating devices. In experiments we show how to route Physarum wires with chemoattractants and electrical fields. We demonstrate that a Physarum wire can be grown on almost bare breadboards and on top of electronic circuits. The Physarum wires can be insulated with a silicone oil without loss of functionality. A Physarum wire self-heals: the ends of a cut wire merge together and restore the conductive pathway in several hours after being cut. The slime mould wires will be used in future designs of self-growing wetware circuits and devices, and integration of slime mould electronics into unconventional biohybrid systems.[3]

The slime mould is capable of sensing tactile,[6,8] chemical[7] and optical[15] stimuli and converting the stimuli to characteristic patterns of its electrical potential oscillations. The electrical responses to stimuli may propagate along protoplasmic tubes for distances exceeding tens of centimetres, like impulses in neural pathways do. A slime mould makes a decision about the propagation direction of its protoplasmic network based on information fusion from thousands of spatially extended protoplasmic loci, similarly to a neuron collecting information from its dendritic tree.[5,16] The analogy is distant yet inspiring. We speculate on whether an alternative — would-be — nervous system can be developed and practically implemented from the slime mould.[9] Based on the analogies between the slime mould and neurons, we demonstrate that the slime mould can play a role of primitive mechanoreceptors, photoreceptors and chemoreceptors; we also show how the Physarum neural pathways develop.

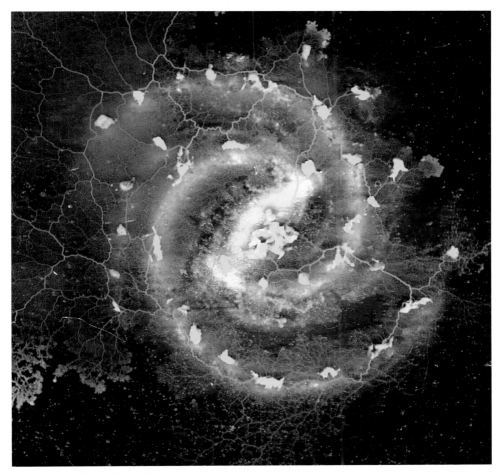

Physarum galaxy. ©2014 Andrew Adamatzky.

Physarum propagating on an artistic impression of a galaxy. See original picture in public domain, NASA/JPL–Caltech.[17] The biological mechanisms underlying the optimal network formation in Physarum machines could be employed in design of large-scale transportation and communication networks in space, where paths between clusters, stars and matter formations are represented by growing protoplasmic tubes.[11]

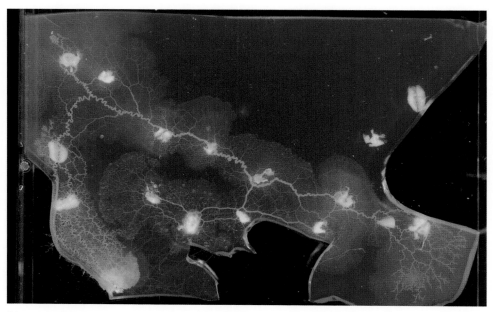

Physarum imitates development of Roman roads in the Balkans. ©2014 Andrew Adamatzky.

We placed oat flakes in the 17 most populated settlement areas of the Balkans during the Roman Empire and a piece of the slime mould in Thessaloniki. In the first 24 hr after inoculation, Physarum propagates from Thessaloniki to Scupi (Skopje) and/or Stobi and Philippoi and from Philippoi to Phillipopolis. In the next 24 hr, Physarum propagates from Serdica, Remesiana–Naissus, Singidunum, Sirmium and Doclea and propagates from Phillipopolis to Hadrianopolis and from Hadrianopolis to Heraclea and Constantinople/Byzantium. By 72 hr of the experiment the slime mould propagates from Doclea to Dyrrhachium and from Dyrrhachium to Nicopolis. Within four days Physarum colonises all cities: the slime mould propagates from Sirmium to Heraclea and Marcianopolis and connects Nicopolis to Scupi (Skopje) and/or Stobi.[18]

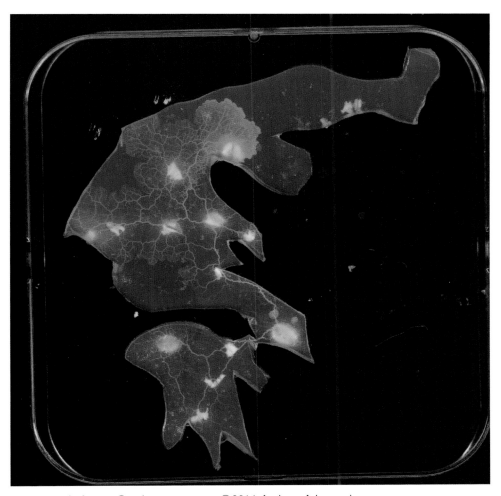

Physarum imitates Greek motorways. ©2014 Andrew Adamatzky.

The slime mould approximates the human-made motorways of Greece. In almost every experiment, Physarum reconstructed the Egnatia road. In half of the experiments, the plasmodium matched the main motorway of Greece, PATHE, and the roads that are currently under construction. The plasmodium foresees the construction of the Ionian motorway — a road connecting the east coast cities of Greece.[19]

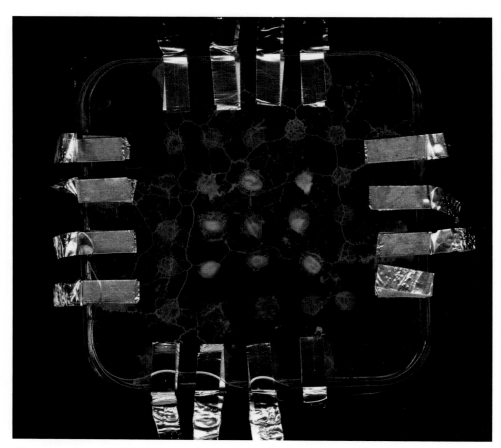

Physarum chip. ©2014 Andrew Adamatzky.
We evolve Physarum computing circuits by selectively tuning topology of protoplasmic networks with electrical current. The Physarum networks can be dynamically controlled via transposon-inspired mechanisms.[20,21]

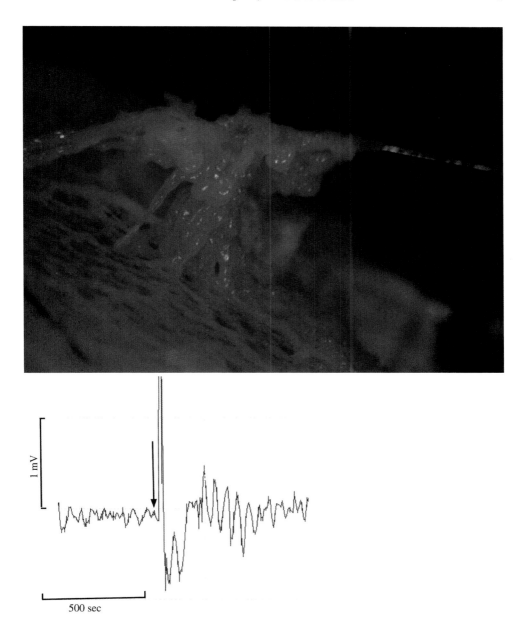

Slime mould tactile bristle. ©2014 Andrew Adamatzky.

Physarum responds to mechanical stimulation, twisting and stretching by unique patterns of oscillations of its membrane potential. Top: photograph of experimental setup of Physarum tactile sensor: a bristle is partly colonised by slime mould. Bottom: when the bristle is deflected, the slime mould changes its pattern of electrical oscillations, thus signalling about the tactile stimulation event.[6]

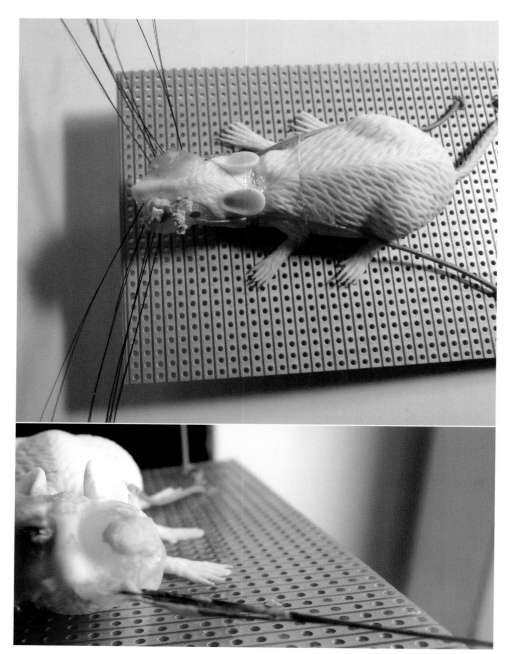

Slime mould tactile bristle. ©2014 Andrew Adamatzky.

Rubber mouse equipped with Physarum bristles: the bristles are partly colonised by Physarum.[6] In experiments with the mouse, protoplasmic tubes connecting agar blobs in the base of the whiskers sometimes propagated across the mouse's nose bridge but sometimes across the inframaxillary region. In most experiments, the bristles were partly colonised by Physarum. The degree of colonisation varied from one-seventh of a bristle to almost half of the bristle length.[6]

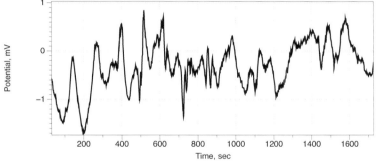

Physarum ocular pathway. ©2014 Andrew Adamatzky.

The slime mould is capable of sensing tactile, chemical and optical stimuli and converting the stimuli to characteristic patterns of its electrical potential oscillations. The electrical responses to stimuli may propagate along protoplasmic tubes for distances exceeding tens of centimetres, like impulses in neural pathways do. We speculate on whether an alternative — would-be — nervous system can be developed and practically implemented from the slime mould.[9] The figure illustrates a functionality test of the Physarum ocular pathway. Top: Physarum colonising a reference electrode in the right eye socket is illuminated with a 1400 lux white spot light. Bottom: electrical activity of Physarum recorded on an electrode based inside the cranium. Light was switched on for 600 sec of recording.[9]

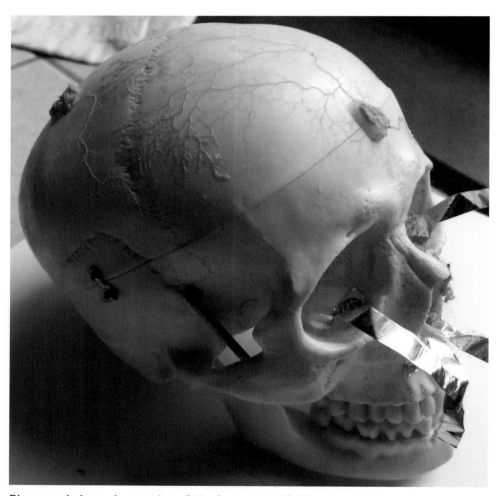

Physarum imitates innervation of the front scalp. ©2014 Andrew Adamatzky.

To imitate sensorial innervation of the front scalp we inoculated Physarum on the frontal bone of a life-size human skull model 50 mm above the glabella and placed a few oat flakes on the parietal bone. In two days Physarum developed an extensively branching tree of protoplasmic tubes. The tree spanned a substantial part of the frontal lobe, even covering its lateral parts, crossed the coronal suture and developed actively branching growing zones moving towards the target site on the parietal bone.[9]

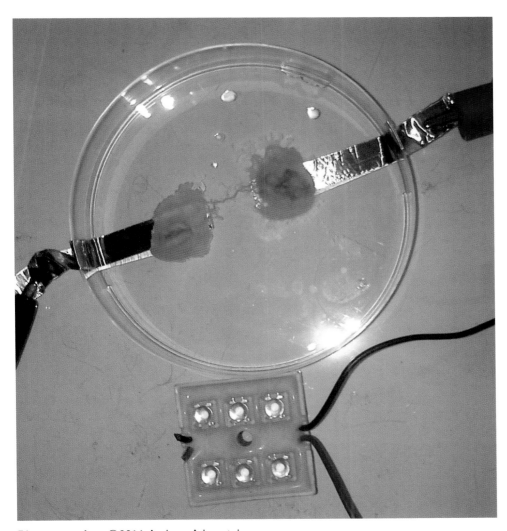

Physarum wire. ©2014 Andrew Adamatzky.

Protoplasmic tubes of Physarum can play a role of self-routing and self-repairing wires. We have conducted experiments on incorporating a living Physarum wire into a circuit which includes an array of six LEDs (15 V white 10,000 Mcd). We applied approx. 19 V to the circuit until the LED array was shining brightly. The potential on the LED array was 4.4–4.8 V and the current 11–13 μA. In all experiments, the Physarum wire was functioning for 24 hr without loss of integrity under load. Top: overall setup. Bottom: Physarum wire 24 hr after functioning in the circuit with LED array.[11]

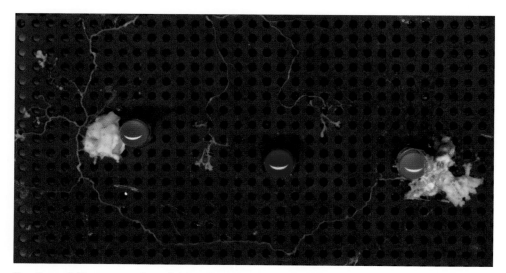

Routing of Physarum wire. ©2014 Andrew Adamatzky.

Physarum wires are routed with attractants and repellents. In the experiment illustrated, we aimed to route a Physarum wire from a position of a yellow LED Y to a position of a green LED G, but avoiding a position of a red LED R. Physarum was inoculated near Y. An oat flake was placed east of G. Chemoattractants released either by the oat flake or by bacteria colonising it diffused in the air and attracted Physarum. To prevent the slime mould going near R, we placed a grain of salt near R. The salt absorbed water from the humid environment of the experimental setup and diffused in the agar layer underlying the breadboard. Physarum is repelled by a high concentration of salt in the substrate. Thus, the Physarum moved towards attracting G and, at the same time, avoided repelling R.[11]

Physarum wire on electronic board. ©2014 Andrew Adamatzky.

To evaluate how well Physarum propagates on a bare surface of electronic components and assemblies, we conducted experiments with electronic boards. Typically, a slime mould was inoculated at one edge of the board, and oat flakes were sparsely distributed on the board to attract Physarum to certain domains of the board. The oat flakes generated chemoattractive fields to guide Physarum wires towards imaginary pins. The boards were kept in a container, with shallow water on the bottom to keep the humidity very high. The boards with Physarum were not in direct contact with the water. The slime mould propagated on the boards with a speed of between 1 mm and 5 mm per hour. Physarum propagated satisfactorily on both sides of the boards, and usually spanned a planar set of oat flakes with networks of protoplasmic tubes ranging from spanning trees to their closures into β-skeletons. The width of protoplasmic wires grown by Physarum is comparable with the width of conductive pathways on the computer boards.[11]

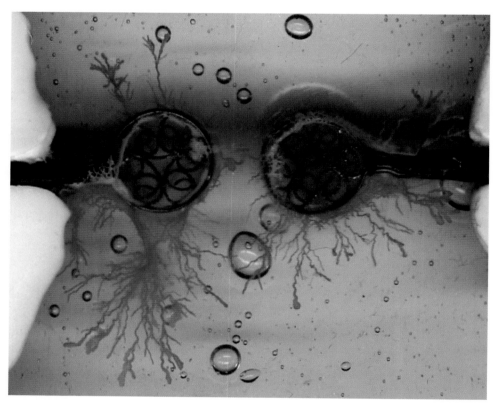

Insulated Physarum wires. ©2014 Andrew Adamatzky.

Physarum wires can be insulated with octamethylcyclotetrasiloxane (D4). The D4 is poured into a container with the slime mould. Physarum becomes completely encapsulated in the silicon insulator. When a slab of D4 is removed from the container, Physarum remains inside. Physarum survived inside D4 for days. The figure shows that the Physarum wire, connecting two probes, is covered by a layer of D4; view from the bottom of the Petri dish.[11]

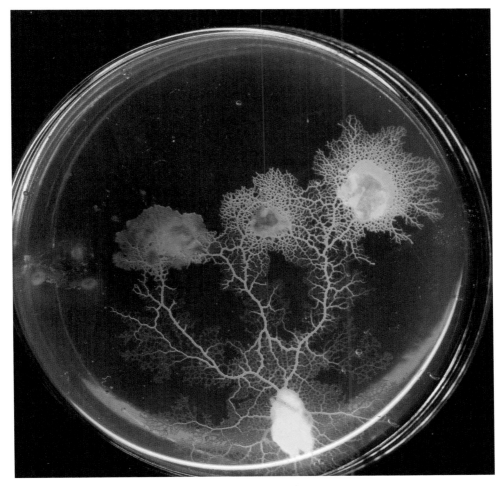

Physarum transports dyes. ©2014 Andrew Adamatzky.

In laboratory experiments we demonstrated that plasmodium of P. polycephalum consumes food colourings and distributes them in its protoplasmic network. By specifically arranging a configuration of attractive — sources of nutrients — and repelling — increased concentration of sodium chloride — fields, we program the plasmodium to implement the following operations: to take in colouring from the closest coloured oat flake; to mix two different colours to produce a third colour; and to transport the colour to a specified locus of the experimental substrate. The plasmodium demonstrates intelligent behaviour by avoiding "hazardous" domains of the environment when propagating towards target colours and by minimising the length of the transport route between its source of origin and the target colour. These findings manifest the plasmodium's potential for being a primary component for amorphous biorobotic devices. A plasmodium biorobotic device transports substances by pumping the substance through the plasmodium's protoplasmic network and by direct relocation, or migration, of the whole plasmodium's body.[22]

Localised Physarum propagates on agar gel. ©2014 Andrew Adamatzky.

This is an example of how Physarum forms dissipative soliton-like structures. The slime mould active zone exhibits a characteristic wave front with a tail of protoplasmic tubes trailing behind. The active zone resembles wave fragments (dissipative solitons) in a Belousov–Zhabotinsky medium. The active zone is an elementary processor of a multi-processor Physarum machine.[11,23,5]

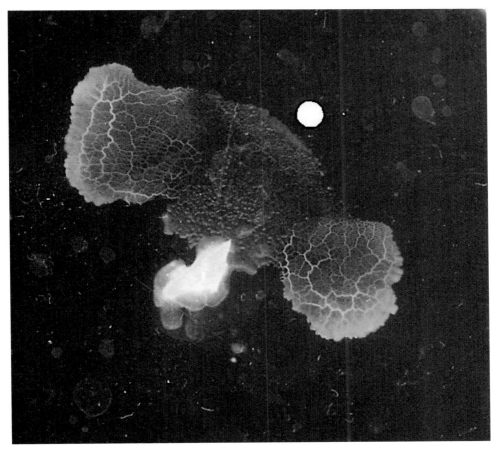

Controlling Physarum with repellents. ©2014 Andrew Adamatzky.
Physarum wave fragment travelling north-east "collides" with a grain of salt (white disc) and splits into two independent fragments; one fragment travels north-west and the other south-east.[11,5]

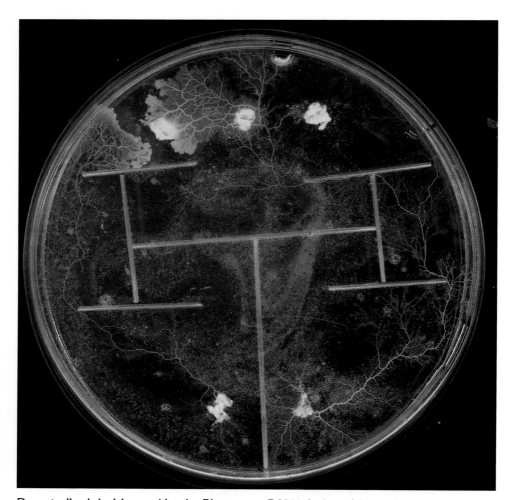

Decentralised decision making by Physarum. ©2014 Andrew Adamatzky.
Oat flakes with plasmodium were placed in the south part of a Petri dish. Virgin oat flakes were placed in the north part. Obstacles were represented by capillary tubes placed on an agar surface. Optimal — from the Physarum machine's point of view — paths connecting source and destination sites are seen as pronounced protoplasmic tubes.[11]

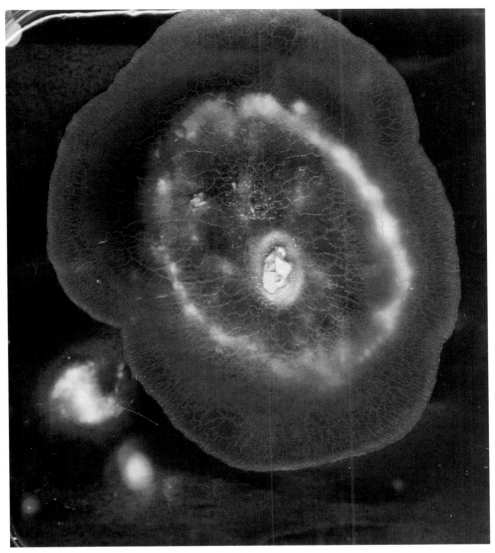

Physarum imitates expanding galaxy. ©2014 Andrew Adamatzky.

Image of Physarum growing on a nutrient agar superimposed on a false-colour composite image of the Cartwheel Galaxy: see PIA03296: A Stellar Ripple, NASA/JPL–Caltech. On a nutrient substrate Physarum expands as an omnidirectional wave, e.g. as a classical excitation wave in a two-dimensional excitable medium. It shows a pronounced wave front, comprising a very dense network of protoplasmic tubes. Morphological transitions of the slime mould's networks during expansion, colonisation and development bear a remarkable resemblance to the Cosmic Web.[11]

Physarum spreads out of enclosure. ©2014 Andrew Adamatzky.

Propagation of Physarum on a nutrient agar gel with only partial geometrical constraining. Obstacles are represented by rectangular shapes of plastic inserted into the agar layer. Physarum spreads out of the enclosure as a wave in an active medium.[4]

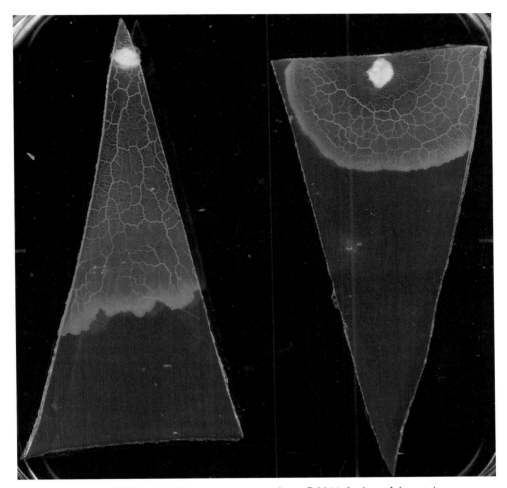

Geometry-induced delays in Physarum propagation. ©2014 Andrew Adamatzky.

Given a triangle cut out of a nutrient agar plate and Physarum inoculated either at the vertex of the triangle or in the middle of its base, Physarum propagates almost one and a half times quicker from the triangle's top to the base than from the base to the top.[4]

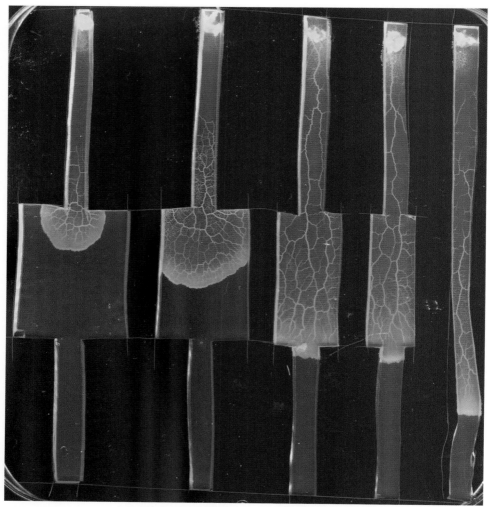

Geometry-induced delays in Physarum propagation. ©2014 Andrew Adamatzky.
To implement a delay element in a Physarum circuit, one can add a fixed size expansion to a propagation channel. Physarum delays propagation on a stripe with expansion width w by $0.14 \times \ln(w) + 0.9$ times compared to the Physarum's propagation on a stripe without expansion.[4]

Physarum imitates migration in USA. ©2014 Andrew Adamatzky.

The slime mould is a living analogue model of human migration at a large scale. We found that the protoplasmic network is a good biological model of a migration network emerging in a transnational community of people living between Mexico and the USA. In a laboratory experiment with the slime mould, we imitated Mexican migration on a three-dimensional nylon model of the USA. Migratory links developed on a flat substrate may represent air transportation while routes on three-dimensional terrain signify ground transportation.[24]

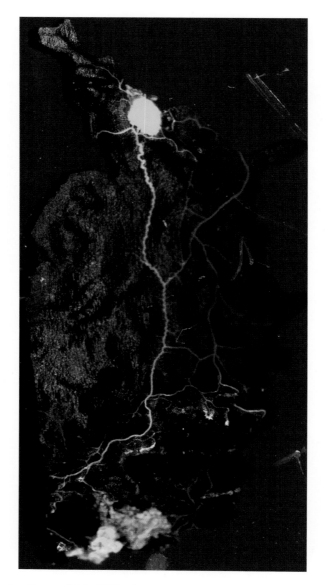

Physarum imitates Route 20 in USA. ©2014 Andrew Adamatzky.

To imitate Route 20 in the USA, we inoculate slime mould in Newport, Oregon, and place fertile oat flakes, to attract the slime mould, in Boston, Massachusetts. It takes slime mould 4–5 days to cover the distance between Newport and Boston. In experiments with the USA, one day of slime mould's propagation roughly corresponds to about 10 hr driving along Route 20. On the three-dimensional terrain of the USA the average slime mould route is 1.095 times longer than Route 20.[25]

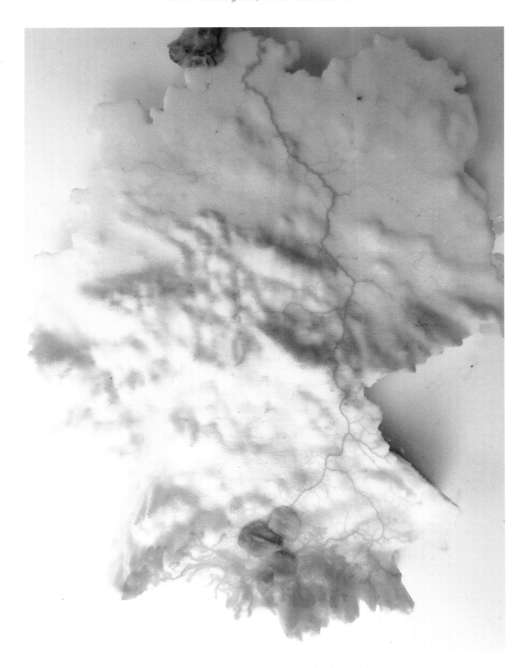

Physarum imitates Autobahn 7 in Germany. ©2014 Andrew Adamatzky.

To imitate Autobahn 7 in Germany, we inoculate slime mould in Flensburg and place oat flakes, to attract the slime mould, in Füssen. It takes slime mould 2–3 days to propagate from Flensburg to Füssen. In experiments with Germany, one day of slime mould's propagation roughly corresponds to 3–5 hr of real-life driving along Autobahn 7. On the terrain of Germany, the average slime mould route is 1.158 times longer.[25]

Physarum navigates around mountains in USA. ©2014 Andrew Adamatzky. Configuration of a path developed is determined by a level of activity of an imitating plasmodium, expressed via the relation between minimum and maximum allowed levels of excitation and maximum height of elevations it is allowed to climb on.[25]

Physarum navigates around mountains in Germany. ©2014 Andrew Adamatzky.
Absence of any guidance force in a substrate together with non-flat terrain allows the slime mould to deviate, sometimes substantially, from a shortest route towards its destination site.[25]

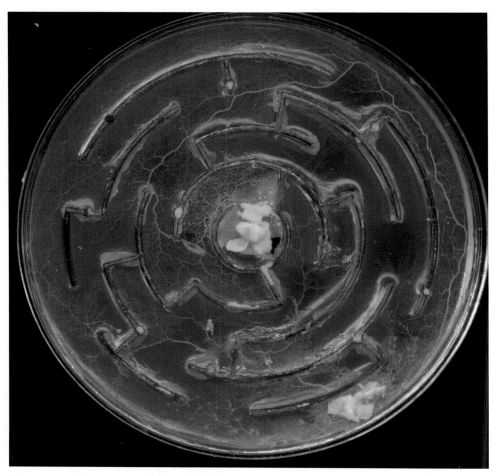

Physarum solves maze. ©2014 Andrew Adamatzky.

If plasmodium of Physarum is inoculated in a maze's peripheral channel and an oat flake (source of attractants) in the maze's central chamber, then the plasmodium grows towards the target oat flake and connects the flake with the site of original inoculation with a pronounced protoplasmic tube. The protoplasmic tube represents a path in the maze. The plasmodium solves the maze in one pass because it is assisted by a gradient of chemoattractants propagating from the target oat flake.[14]

Physarum imitates World colonisation. ©2014 Andrew Adamatzky.

To imitate a hypothetical colonisation of the World and formation of major transportation routes, we cut continents on the surface of a three-dimensional globe covered by non-nutrient agar, represent positions of selected metropolitan areas with oat flakes and inoculate the plasmodium in one of the metropolitan areas. The plasmodium propagates towards the sources of nutrients, spans them with its network of protoplasmic tubes and even crosses the bare substrate between the continents. From the laboratory experiments we derive weighted Physarum graphs, analyse their structure and compare them with the basic proximity graphs and generalised graphs derived from the Silk Road and the Asia Highway networks.[26]

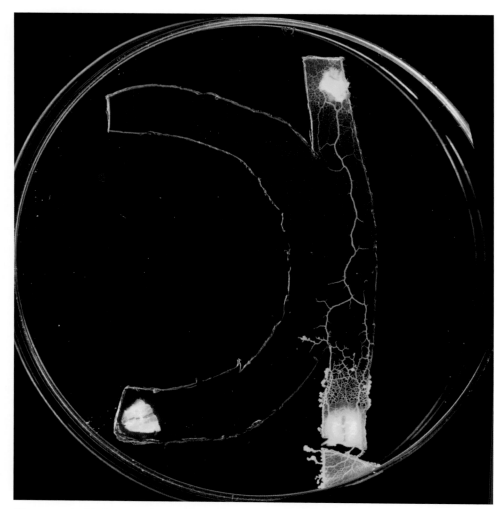

Physarum ballistic gate. Inputs 0 and 1. ©2014 Andrew Adamatzky.
On a non-nutrient substrate the plasmodium propagates as a travelling localisation, as a compact wave fragment of protoplasm. The plasmodium localisation travels in its originally predetermined direction for a substantial period of time even when no gradient of chemoattractants is present. We utilise this property of Physarum localisations to design a two-input, two-output Boolean logic gate $\langle x, y \rangle \rightarrow \langle xy, x + y \rangle$. Experimental example of transformation $\langle x, y \rangle \rightarrow \langle p, q \rangle$ implemented by Physarum ballistic gate: $\langle 0, 1 \rangle \rightarrow \langle 0, 1 \rangle$.[2]

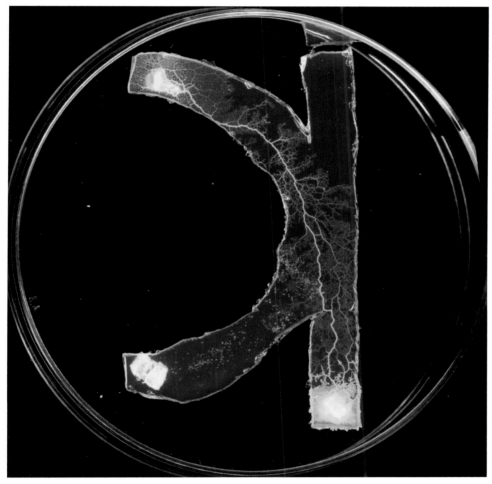

Physarum ballistic gate. Inputs 1 and 0. ©2014 Andrew Adamatzky.

Experimental example of transformation $\langle x, y \rangle \to \langle xy, x+y \rangle$ implemented by Physarum ballistic gate: $\langle 1, 0 \rangle \to \langle 0, 1 \rangle$.[2]

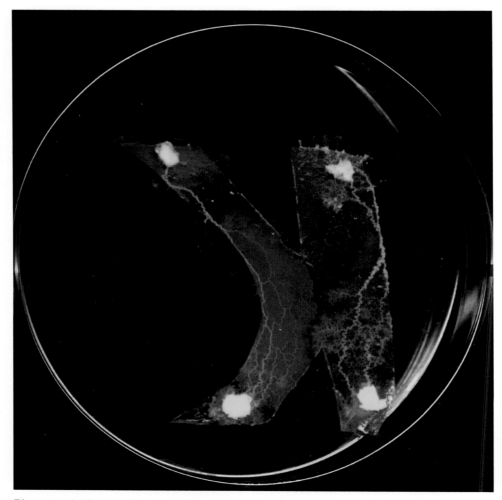

Physarum ballistic gate. Inputs 1 and 1. ©2014 Andrew Adamatzky.
Experimental example of transformation $\langle x, y \rangle \to \langle xy, x+y \rangle$ implemented by Physarum ballistic gate: $\langle 1, 1 \rangle \to \langle 1, 1 \rangle$.[2]

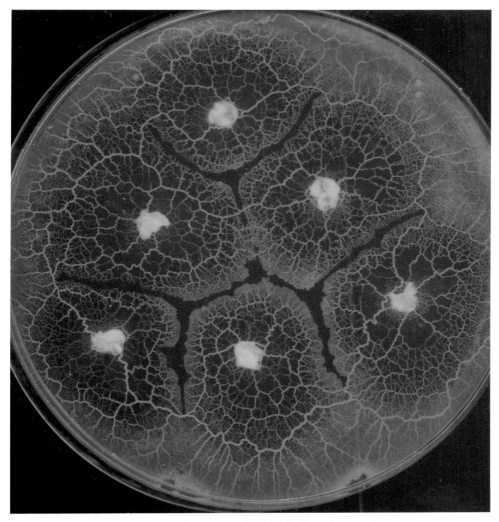

Physarum approximates Voronoi diagram. ©2014 Andrew Adamatzky.

A planar Voronoi diagram (VD) of a set **P** is a partition of the plane into such regions that, for any element of **P**, a region corresponding to a unique point $p \in$ **P** contains all those points of the plane which are closer to p than to any other node of **P**. The plasmodium growing on a nutrient substrate from a single site of inoculation expands circularly as a typical diffusive or excitation wave. When two plasmodium waves encounter each other, they stop propagating. To approximate a VD with Physarum, we physically map a configuration of planar data points by inoculating plasmodia on a substrate. Plasmodium waves propagate circularly from each data point and stop when they collide with each other. Thus, the plasmodium waves approximate a VD, whose edges are the substrate's loci not occupied by plasmodia.[5]

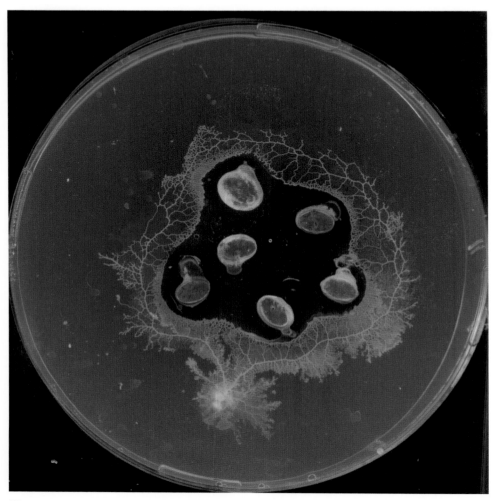

Physarum approximates hull of a planar set. ©2014 Andrew Adamatzky.
Computing a polygon defining a set of planar points is a classical problem of modern computational geometry. In laboratory experiments, we demonstrated that a concave hull, a connected α-shape without holes, of a finite planar set is approximated by the slime mould *P. polycephalum*. We represent planar points with sources of long-distance attractants and short-distance repellents and inoculate a piece of plasmodium outside the data set. The plasmodium moves towards the data and envelops it by pronounced protoplasmic tubes.[27]

Chapter 2

Multi-agent model of slime mould for computing and robotics

Jeff Jones

Unconventional Computing Centre,
University of the West of England, Bristol, United Kingdom
jeff.jones@uwe.ac.uk

This chapter features results from a multi-agent model of slime mould. Slime mould is a remarkable organism because it possesses no nervous system, no skeleton, no organised musculature and no special senses. Despite these limitations, slime mould is capable of remarkable biological and computational feats by dynamically adapting its body plan in response to environmental stimuli. Because slime mould consists of simple component parts, its behaviour requires no specialised or critical components and the mechanisms which govern its behaviour are distributed throughout — and embedded within — the organism itself. This multi-agent approach to modelling slime mould is a bottom-up model and attempts to specifically use the same — apparently limiting — properties found in the organism itself: simple component parts, local interactions and self-organised collective emergent behaviour.[28] The aim of the model is to show how the complexity of slime mould can emerge from these very simple local interactions. The model has successfully been applied to reproducing the biological behaviour of slime mould (growth patterns, network adaptation, oscillatory phenomena) and also the computational and robotic behaviour of slime mould. The images in this chapter give a flavour of the model with topics relating to the complex pattern formation phenomena, amoeboid movement and collective transport phenomena, and its utilisation as a spatially represented unconventional computing substrate.

The first image demonstrates how the model may be considered as a reaction–diffusion pattern formation mechanism where local activation phenomena (movement of the agent particles forms trails which serve to further attract nearby particles) and lateral inhibition phenomena (depletion of the uniform distribution of agents as they move) interact.[28] The second image demonstrates the fundamental basis of slime mould's behaviour: the self-assembly and evolution of transport networks from a solid "sheet" of virtual material. We then demonstrate, in the third image, the complex Physarum-like growth of the model plasmodium when inoculated in the centre of its nutrient-rich environment.

The ability of slime mould to propel itself through its environment is attractive to robotics because both the impetus and the control of the amoeboid movement are distributed within the material of which it is composed. This is evidenced by the fact that slime mould can even be cleaved into two or more separate independently foraging organisms and also re-fused together to construct a larger organism. Such distributed material robotics would be very desirable and we demonstrate images which show the emergence of amoeboid movement in the model in which a virtual blob of slime mould moves via the emergence of internal travelling waves towards a source of nutrients. The potential robotics applications are shown by the movement of the blob through a narrow grating, where the blob rearranges its body plan to fit through the grating, before reassembling itself as a uniform blob.[29] The potential for collective transport is explored in the next image in which travelling waves in a fixed patch of virtual plasmodium material are utilised in simulation to transport simple objects by means of simple ciliate-like movements.[30]

The final images in this chapter explore the potential to employ the model as a spatially extended unconventional computing substrate. We demonstrate the approximation of combinatorial optimisation problems,[31] two-dimensional geometry problems[32] and path-planning problems.[33] The model demonstrates how we can learn from simple organisms such as slime mould to develop techniques which exploit the power of self-organisation between simple components to produce extremely complex behaviours.

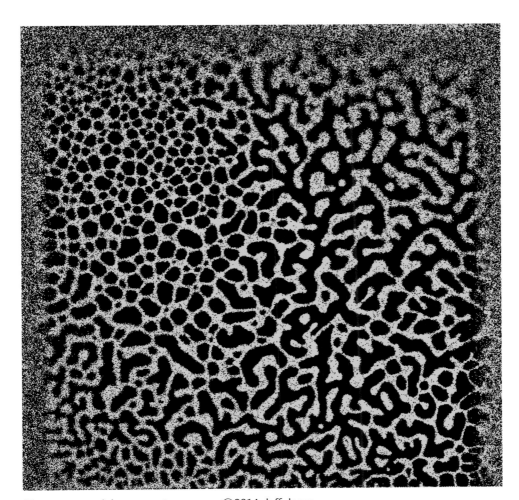

Physarum models parameter space. ©2014 Jeff Jones.

Parametric map of multi-agent model of Physarum patterns. X-axis represents the agent sensor angle parameter, Y-axis represents the agent rotation angle parameter. The range of both parameters is 0–180°. Lattice size was 720 × 720 pixels with 181,440 particles. A complex reaction–diffusion patterning is generated.[28]

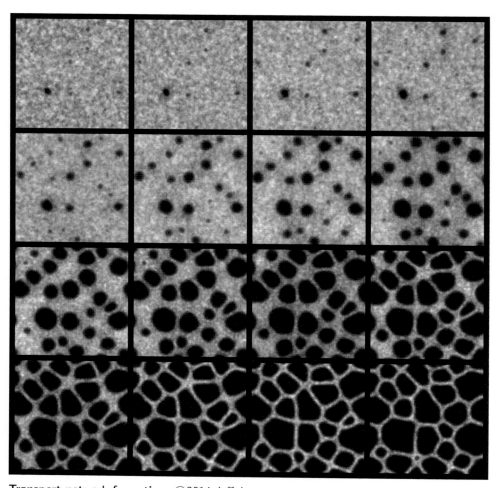

Transport network formation. ©2014 Jeff Jones.

Self-organised formation of transport network from a solid sheet of virtual plasmodium. From an initially disordered flux of particles comprising the virtual plasmodium, pores begin to form as particles are randomly removed. These pores grow in size as the population size further reduces. The edges between the pores form into minimising transport networks.

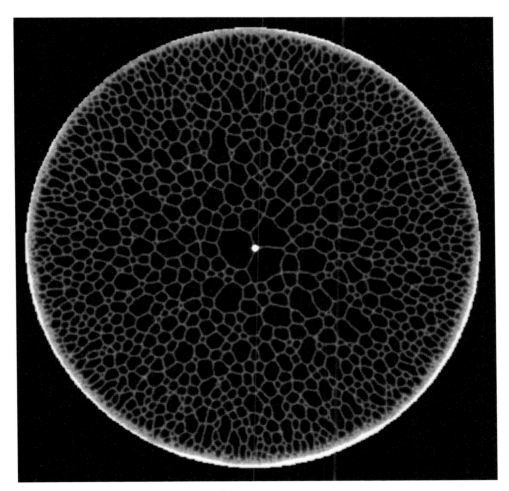

Physarum models growth. ©2014 Jeff Jones.

Growth of virtual slime mould transport network from a central inoculation site on a simulated nutrient-rich substrate. The morphologies of the model networks are strongly affected by substrate concentration.[28]

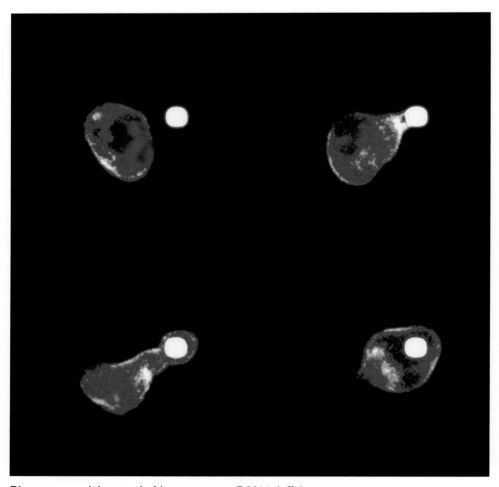

Physarum models amoeboid movement. ©2014 Jeff Jones.
Amoeboid movement in virtual slime mould as it moves towards and engulfs a nutrient stimulus. The movement occurs by means of emergent travelling waves (brighter regions) which move towards the diffusing food stimulus, shifting the bulk of the model collective.[29]

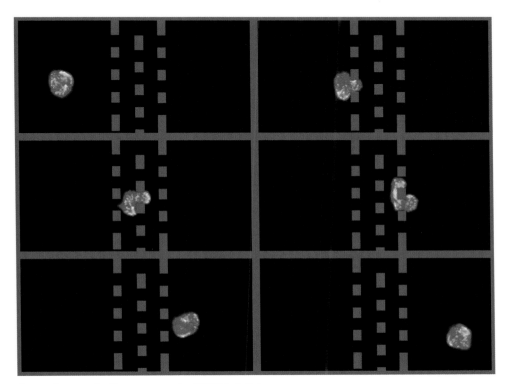

Amorphous robotic movement. ©2014 Jeff Jones.

Collective amoeboid movement in virtual slime mould can be guided by "pushing" the blob with simulated light irradiation. The blob adapts its shape to obstacles (purple blocks) and can pass through spaces which are smaller than the blob itself, before re-establishing its original shape.[29]

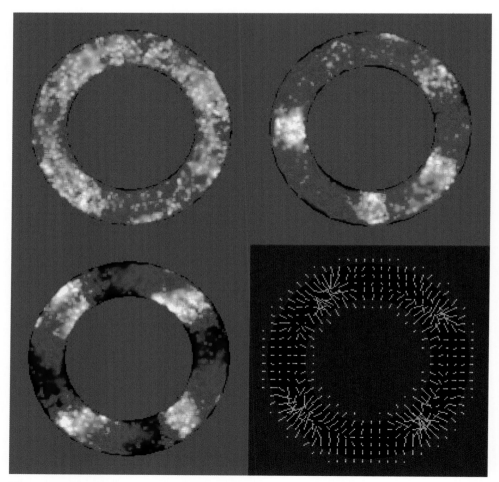

Physarum models engine. ©2014 Jeff Jones.
Collective movement in virtual slime mould can generate regular movement impulses. After initialisation the formation of travelling waves is observed (brighter regions). These waves self-organise into regular circular pulses and can be used to generate collective ciliated movement to transport simulated objects (lower right).[30]

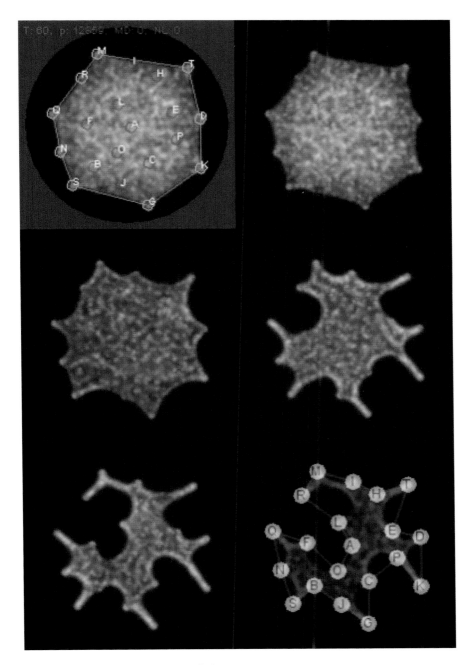

Shrinking blob computer. ©2014 Jeff Jones.

Visualisation of the shrinking blob method to approximate the travelling salesman problem (TSP). A sheet of virtual material is initialised within the confines of the convex hull of a set of cities. As the blob shrinks, it is constrained by the stimulus from the cities and adapts its shape. When all cities are partially uncovered, a tour of the TSP is found by tracing the periphery.[31]

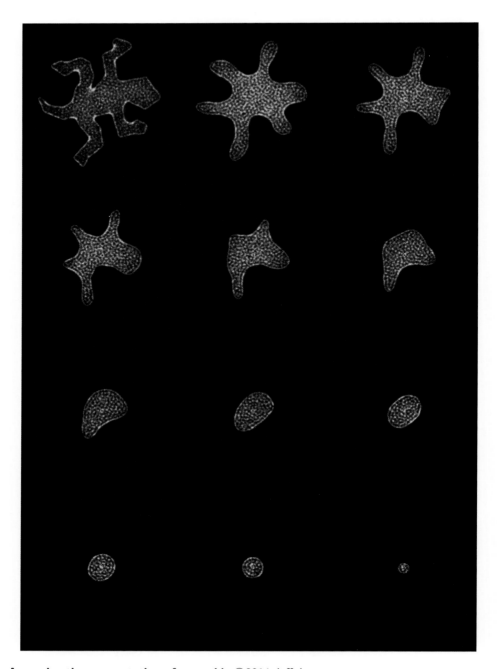

Approximating computation of centroid. ©2014 Jeff Jones.
Shrinkage of virtual plasmodium approximates the geometric computation of the centroid. When inoculated as a two-dimensional planar shape, a shrinking blob of virtual plasmodium withdraws its external projections and forms a minimising shape. By tracking the mean position of the particles, it was found that the shrinking blob approximated the centre of mass of the original two-dimensional shape at the end of the shrinkage process.[32]

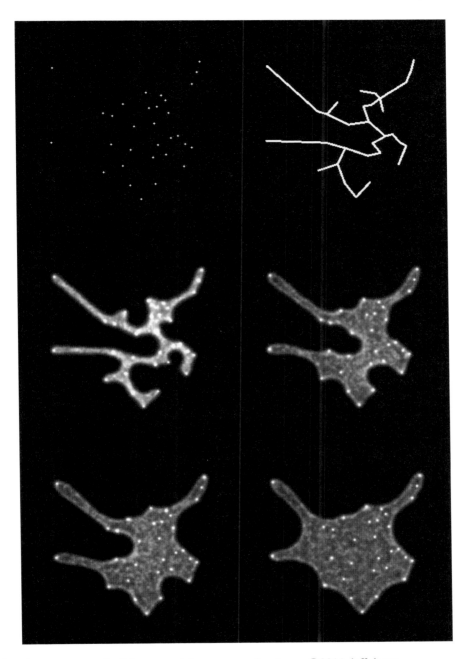

Growing a concave hull from a minimum spanning tree. ©2014 Jeff Jones.

The concave hull represents the "footprint" of a set of points. This example shows the location of major cities in the People's Republic of China (top left). By inoculating the virtual plasmodium on the minimum spanning tree (top right), the virtual plasmodium grows to approximate the concave hull, halting its growth automatically.

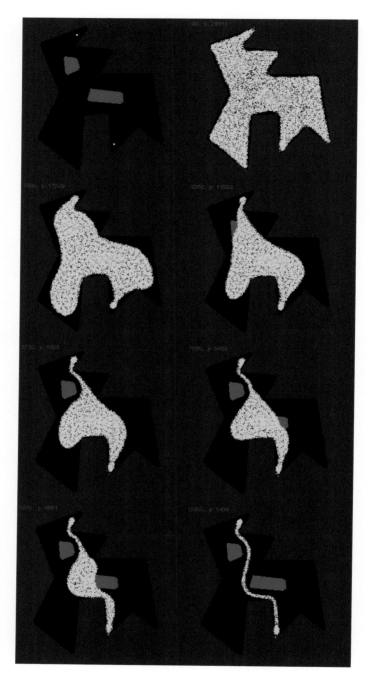

Computation of collision-free path by shrinkage. ©2014 Jeff Jones.
Unconventional computation of collision-free path is performed by shrinking a virtual plasmodium in a two-dimensional arena. The virtual plasmodium is attracted to the start and end points (white dots in top-left image). The shrinking blob adapts its shape to the confines of the arena and avoids the obstacles, which deposit a repellent stimulus.[33]

Chapter 3

Slime mould biotechnology

Richard Mayne

Unconventional Computing Centre,
University of the West of England, Bristol, United Kingdom
richard.mayne@uwe.ac.uk

Slime mould computing is an inherently multi-disciplinary subfield of unconventional computing that draws upon aspects of not only theoretical computer science and electronics, but also the natural sciences. This chapter focuses on the biology of slime moulds and expounds the viewpoint that a deep, intuitive understanding of slime mould life processes is a fundamental requirement for understanding — and, hence, harnessing — the incredible behaviour patterns we may characterise as "computation".

In demonstration of this point, consider the *P. polycephalum* plasmodium navigating its way through a maze. It is tempting to compare this behaviour to that of a robot which has been programmed to complete the same task. Distributed sensing, actuation and decision making have long since been realised *in silico* — indeed, whilst the physical components used in these robots have been in production for several decades, algorithms for solving maze-type puzzles have been around since the 1880s[34] — and hence our perception of these processes is distinctly biased towards these systems. In reality, however, the computational processes that occur within an organism are so divergent from their artificial equivalent that any comparisons between the two are, at best, unhelpful. For example, where an automaton will gather data from an array of sensors that have converted external stimuli into binary electrical signals, slime moulds possess a vastly greater number of microscopic membrane-bound receptors which serve an analogous function.

These receptors constantly survey the environment; multiple varieties are present to respond to a wide range of stimuli, including light, mechanical force, chemicals and temperature. Receptors typically respond to stimulation by instigating one of a range of chemical, electrical or mechanical events: these may be interpreted loosely as forms of "data" which carry signals through the plasmodium. The signals generated by receptors are usually second-messenger systems that trigger a non-linear cascade of consequent effects which eventually alter the cell's metabolism and/or gene expression. For a thorough review of this topic, please see Ref.[35]

Despite the fundamental differences between them, biological and artificial entities can both compute the solutions to common problems, such as logic and arithmetic. Intriguingly, biological entities are superior at some forms of computation than traditional computers: for example, a desktop computer will calculate the solution to, say, a difficult sum far more quickly than a human, but cannot simulate intuitive understanding of natural languages in the way that the human brain can. This highlights how differences in operation define the capabilities of the entity: if we are to harness these novel abilities exhibited by unconventional computing substrates, therefore, we must reinvent the computing paradigm for each new substrate. By extension, if we are to "program" our biological computers, our understanding of them must be based on the science of their functionality.

The images in this chapter detail experiments designed to enhance our understanding of the biology of slime mould, the conclusions drawn from which have been used in the design of slime mould computing devices.

The first three images are confocal laser scanning micrographs that were produced in efforts to demonstrate the presence of the plasmodial cytoskeleton, an intracellular protein scaffold that is putatively involved in "information" transduction throughout the organism. Cytoskeletal proteins (actin and tubulin) were visualised via direct immunofluorescence.[35]

The following three images detail an investigation into the ability of slime mould to function as classical electrical logic gates, employing its natural tendency to migrate away from bright lights. In this study, NOT and NAND logical gates were successfully implemented.[36]

The next three images were produced during successful attempts to functionally hybridise slime mould with nanoscale artificial electrical components. It was found that nanoscale metallic particles are internalised by slime mould (via endocytosis and microinjection) and may significantly alter the electrical properties of the organism.[37,38]

The final two images show fine detail of slime mould cellular structure — initially with a light microscope and finally with an electron microscope — and demonstrate the comparative scale of the functional parts of the plasmodium.

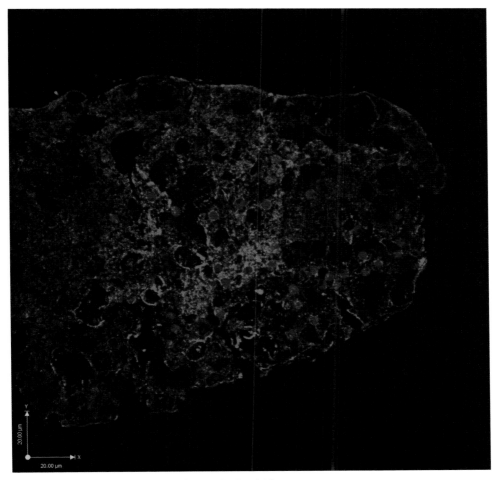

The Physarum actin network. ©2014 Richard Mayne.

Confocal laser scanning micrograph of a 12 μm transverse section through the growing tip of a *P. polycephalum* plasmodium, showing the plasmodial actin network (red) and nuclei (blue). The actin network was stained with an indirect immunofluorescence technique using a monoclonal anti-actin antibody (ACTN05(C4)), and nuclei were stained with a fluorescent nucleic acid analogue, 4′,6-diamidino-2-phenylindole (DAPI). The plasmodial actin network is extremely dense, complex and highly interconnected. Each nucleus appears to articulate onto the actin network in many locations. From images such as these, a model of Physarum computing based on cytoskeletal transduction of data has been created.[35]

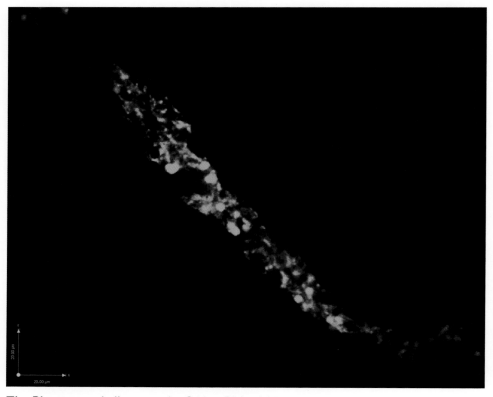

The Physarum tubulin network. ©2014 Richard Mayne.

Confocal laser scanning micrograph of a 4 μm longitudinal section through a minor plasmodial vein/tube structure, showing the plasmodial tubulin network (green) and nuclei (blue). Tubulin was stained with a direct immunofluorescence technique using a monoclonal anti-tubulin antibody (KMX-1), and nuclei were stained with DAPI. As with the plasmodial actin network, tubulin is highly abundant in the *P. polycephalum* plasmodium and seems to connect many of the functional elements of the cell together. Tubulin is another component of the plasmodial data network in our models of Physarum computation.[35]

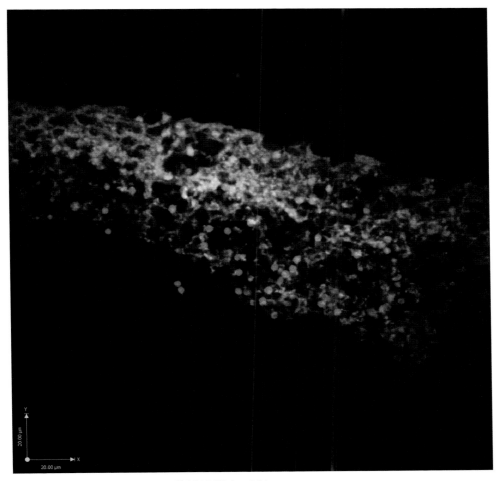

The Physarum cytoskeleton. ©2014 Richard Mayne.
Confocal laser scanning micrograph of a 12 μm transverse section through the growth tip of a *P. polycephalum* plasmodium, stained for cytoskeletal proteins. Actin is red and tubulin is green; both have been visualised by indirect immunofluorescence with anti-actin/tubulin antibodies (ACTN05(C4) and YL1/2, respectively). Nuclei are stained blue with fluorescent nucleic acid dye DAPI. This image demonstrates the complexity, abundance and massive interconnectivity of the plasmodial cytoskeleton, and provides a biomolecular basis for our models of slime mould computation.[35]

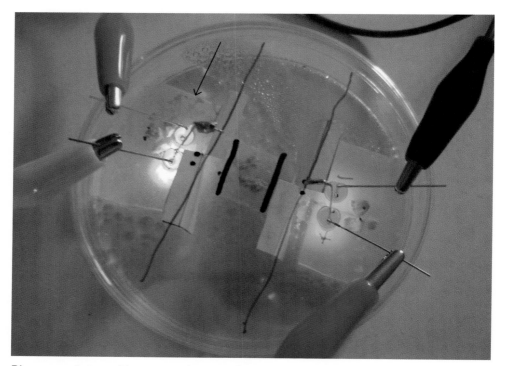

Physarum photoavoidance experiments. ©2014 Richard Mayne.

Photograph of an experiment designed to determine which colour of LED-generated light *P. polycephalum* is most phobic to, based on the principle that slime moulds avoid light. In the experimental environment (a Petri dish), three cubes of 2% non-nutrient agar are present, each loaded with chemoattractants (oat flakes). Slime mould is placed on the central agar piece. The other two agar pieces are located in the left- and right-hand poles of the dish and are illuminated by LED arrays. There is a 20 mm channel separating the left/right agar pieces from the central one and opaque cardboard dividers with 2 mm of base clearance are situated centrally in these channels to prevent light contamination between each array. The slime mould is therefore offered a choice: migrate towards colour A/B or stay still; the latter option implies death as food and moisture sources are limited on the central piece, and so the slime mould rarely chooses this option. By experimenting with different combinations of colours of light, we can determine which colour of light *P. polycephalum* is least phobic to. In this example, the slime mould (arrowed) has chosen to migrate towards the blue-illuminated side rather than the green. The knowledge gained through these experiments has been applied to design slime mould computing devices which utilise LEDs as repellents.[36]

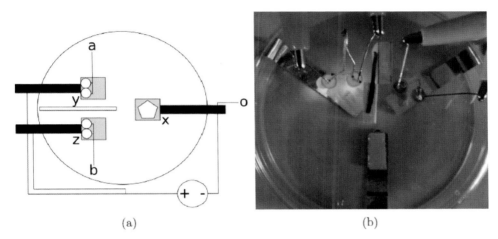

Physarum logical gates. ©2014 Richard Mayne.

(a) Diagram demonstrating the design of a slime mould logical gate. A logical gate is an electronic device which receives multiple electrical inputs and will produce an output if the input configuration matches a predefined pattern. This design exploits slime mould light avoidance and electrical conductivity: slime mould (white pentagon) is placed on a cube of agar (grey squares) overlying an aluminium electrode (black rectangles, labelled x, y, z) in an empty Petri dish. Two other unoccupied blobs/electrodes are present 10 mm away, and all are connected to a power supply. Two LED arrays are mounted in the lid of the Petri dish and are connected to separate power supplies (a and b): these are the inputs.[36] (Continued overleaf.)

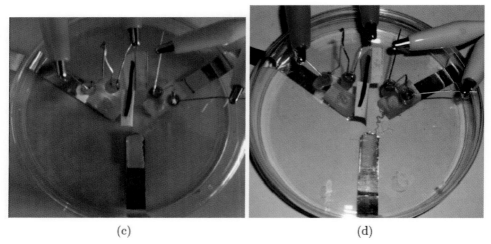

(c)   (d)

Physarum logical gates (continued). ©2014 Richard Mayne.

When an input is live, the LEDs illuminate and prevent the slime mould growing to complete the output circuit between the initial and illuminated electrodes, and *vice versa*. If both electrodes are illuminated, the slime mould retreats, leaving the circuit uncompleted. If neither electrode is illuminated, the slime mould migrates to either of the electrodes and completes the circuit. This device's functionality is the same as a NAND logical gate; this demonstrates how slime mould may be used to carry out logical computation. (b–d) Photographs demonstrating the slime mould NAND gate in action (6 hr time steps), where the left-hand LED array is illuminated.[36]

Microinjecting Physarum with fluorescent nanoparticles. ©2014 Richard Mayne.

Photograph of live *P. polycephalum* plasmodium being microinjected with an aqueous suspension of fluorescent nanoparticles. The slime mould, which is growing in a Petri dish filled with 2% non-nutrient agar, is loaded onto an inverted fluorescence microscope: microinjection — which is performed with a motorised syringe connected to an extremely fine glass needle — is performed with the aid of the microscope, and the newly injected microparticles can be observed shuttling through the organism shortly afterwards with the microscope's fluorescence lamp. These experiments were performed in order to ascertain the best way of introducing artificial circuit elements into slime mould for the production of hybrid artificial organic devices.

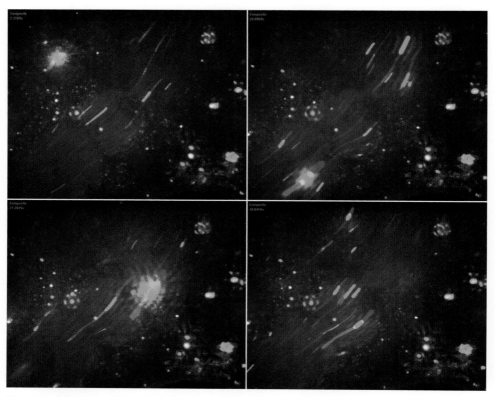

Nanoparticles travelling through Physarum. ©2014 Richard Mayne.
Image gallery of confocal laser scanning microscopy video footage of fluorescent nanostructures within a live *P. polycephalum* plasmodial vein/tube (time steps approx. 1 sec). Red and blue objects represent 200 nm magnetite (iron II/III oxide) nanoparticles conjugated to a fluorophore (perylene), and green objects represent 100 nm fluorescent latex nanospheres. Over time, particles can be seen travelling at great speed (in excess of 100 μm per second) through the hydrodynamic core of this plasmodial tube (the endoplasm, running diagonally from lower left to top right). Several become stuck in the peripheral gel layer of the plasmodial tube (top-left and bottom-right corners). This footage demonstrates how slime mould may transport a range of internalised nanoscale structures and deposit them to its peripheral tissue layers, leaving a coherent trail of particles following plasmodial migration or death. This knowledge allows us to grow biomorphic mineralised circuitry.[38]

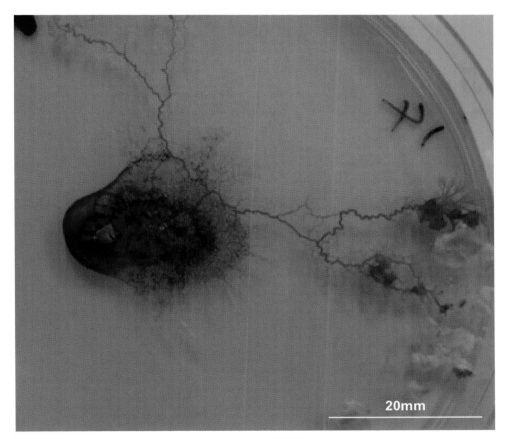

Hybridising Physarum with metallic particles. ©2014 Richard Mayne.

Photograph of a P. polycephalum plasmodium following hybridisation with 200 nm magnetite (iron II/III oxide) nanoparticles. Note how the plasmodium has changed colour, but otherwise looks healthy — i.e. its morphology is normal and it is attempting to forage the oat flakes (right) as usual. Hybridisation was achieved by homogenising the plasmodium with a scalpel blade and dropping a set quantity of aqueous nanoparticle suspension on top.[37]

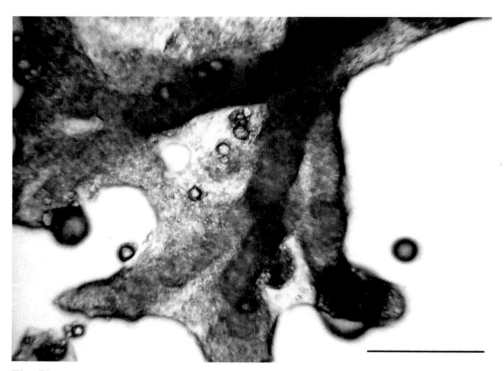

The Physarum growth tip. ©2014 Richard Mayne.
Light micrograph of the growing tip of a *P. polycephalum* plasmodium, using a deep-field technique. *P. polycephalum*'s advancing anterior margin is described as "fan-shaped" when viewed with the naked eye. Microscopically, however, this margin is revealed to be formed from a confluence of multiple fluid channels, similar in appearance to the larger macroscopic plasmodial veins/tubes. The amorphous tips of this region (pseudopodia) are formed by directional assembly of the plasmodial cytoskeleton — a protein scaffold which provides structural support and a certain amount of propulsive force to guide growth of the fluid channels. Scale bar: 100 μm.

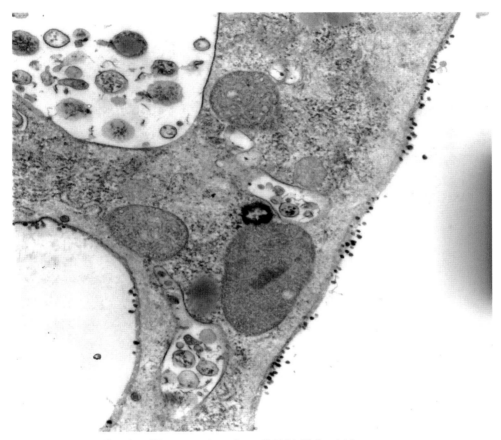

Ultra-high-magnification Physarum imaging. ©2014 Richard Mayne.

Transmission electron micrograph of an ultra-thin (100 nm) transverse section through a plasmodial tube/vein. This micrograph shows the outer membrane (the right hand side portion of the image corresponds with the outside of the organism) of the plasmodium. Several endocytotic vesicles are in the process of being internalised by the organism. These vesicles (white spaces within the organism) contain bacteria and other detritus the organism was feeding on at the time of fixation. Three mitochondria (dark oblongs) are also present. Original magnification: ×52,000.

Chapter 4

Slime mould interactions with chemicals and materials

Benjamin De Lacy Costello

*Institute for Biosensing Technology and Unconventional Computing Centre,
University of the West of England, Bristol, United Kingdom*
Ben.DeLacyCostello@uwe.ac.uk

At first sight a seemingly jumbled selection of images, what links all these together? It is an assessment of how *P. polycephalum* interacts with a synthetic human-made environment and how this can give indications of its natural environmental interactions but more importantly how these can be harnessed to give tangible outcomes in functional material synthesis and biologically inspired computing.

The following excerpts from larger studies outline how *P. polycephalum* can pattern/etch active materials, thus altering their bulk and surface properties. The next illustration outlines how *P. polycephalum* can take up certain chemicals from a substrate and actively transport other chemicals producing an *in situ* reaction and formation of functional materials within the tubular network. These selectively functionalised networks can be utilised as simple interconnects or act as functional electronic components within a circuit.

Next we give an example of *P. polycephalum* as an extremeophile able to survive in harsh chemical environments for long periods of time. The ability of the plasmodial phase to out-compete other microorganisms in these environments gives rise to the longevity. It also highlights that the plasmodial phase is genetically well adapted to survive in both cold and harsh chemical environments, making it an ideal candidate for hybrid system fabrication.

The plasmodial phase adapts its morphology according to the material properties of the substrate; thus, this is a facile method to "program" the morphology, thus potentially obtaining an optimal configuration within a circuit or other computing substrate.

We have mentioned that *P. polycephalum* can out-compete various microorganisms, but it can also act cooperatively. Thus, its use of filamentous fungi as a scaffold for growth highlights how the plasmodial phase can grow over a variety of synthetic three-dimensional structures to form for example smart sensors and actuators.

An example of Physarum growing on a chemical substrate months after inoculation leads to the intriguing possibility that the plasmodium is utilising the mineral product. Certainly it is redistributing the precipitated material in a process of "demineralisation". This is another example of its ability to survive in relatively extreme environments.

The functionalisation of *P. polycephalum* with conducting polymers and other semiconductors is presented. The functionalisation with conducting polymers is interesting because it enables selective pretreatment of a plasmodial tube with a precursor and subsequent exposure to a monomer, giving *in situ* chemical polymerisation. The process leads to highly conducting localised parts of the plasmodial network, whilst other parts are intact.

We present results from a *P. polycephalum* vapour sensor. We are utilising its natural ability to sense volatile chemicals in its environment, alongside its electrical conductivity to detect the electrical resistance change caused by pulses of gas.

The chemotaxis of slime moulds is well studied but we present work which shows that they exhibit chemotactic responses to a range of environmental chemicals which are the products of insect and plant communication. It is appealing to be able to exert control over the growth and morphology of the slime mould simply using volatile plumes. Furthermore, it is shown that this observation can be harnessed to direct signal propagation in simple Physarum-based circuits. The potential for exploiting this phenomenon further in designing new circuits and computing architectures is intriguing.

Patterning surface of active materials. ©2014 Ben de Lacy Costello.

P. polycephalum is capable of patterning active materials including semiconducting materials. Here it is living on precipitated iron ferricyanide. Chemically modified analogues of this material have been suggested as computer memory elements due to their magnetic properties. The chemical reaction pictured is in progress as the precipitation front is still advancing. This demonstrates the ability of Physarum to live in extreme environments. In this case, it also shows its adaptability; it is able to avoid the high reactant concentrations, but repopulate areas where the product is formed. This behaviour could be useful in producing active materials via a Physarum-mediated "lithographic" process. The important factor is to be able to control network formation and morphology.

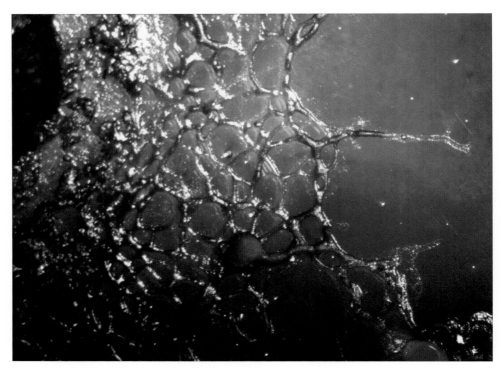

In-tube formation and transport of active materials. ©2014 Ben de Lacy Costello.

P. polycephalum living on a substrate loaded gel (such as potassium ferricyanide) at an appropriate concentration (i.e. which is not directly toxic and does not inhibit network growth) will actively transport the substrate throughout its existing network structure. If a reactant solution is subsequently added to the source of the Physarum network (usually the inoculation source), then this is also actively transported and the chemical reaction occurs *in situ*. This active chemical process is one method of obtaining a functionalised network. This method can be employed to obtain a range of electrochemically active and semiconductor-based materials. The image shows the active deposition of these materials throughout the network structure of Physarum. At the network terminal ends, it is possible to observe the active nature of the transport process, which is being inhibited by the chemical reaction (in the majority of cases).

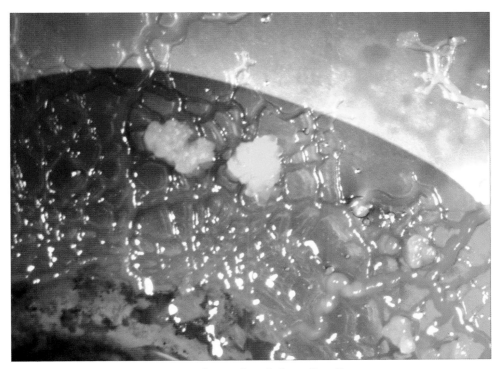

P. polycephalum—extremeophile. ©2014 Ben de Lacy Costello.

Physarum shown here seems very happy growing over iron-based precipitates, in the presence of the primary reactants ferrous sulphate and potassium ferricyanide. Thus, Physarum with its tolerance for extreme chemical environments potentially lends itself to use in bioinorganic hybrid materials for sensing and computing applications. It can also survive in and is adapted for other extreme conditions such as cold and zero/microgravity, thus making it an ideal organism for space exploration and other extreme environments.

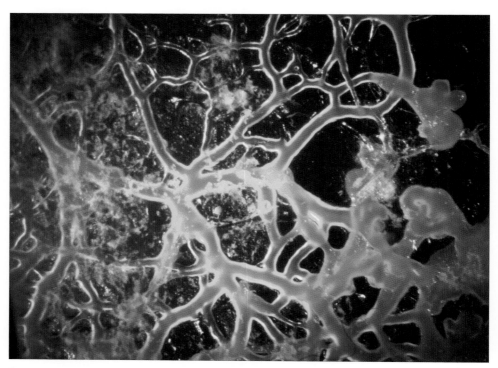

Substrate-dependent morphology. ©2014 Ben de Lacy Costello.

P. polycephalum adapts its morphology according to the substrate it is growing on. In this case the image shows network formation on a polymethylmethacrylate surface. The networks formed tend to be simpler with larger tubes and nodes and less branching, presumably a physical effect mediated by surface adhesion and interfacial surface tension. Thus, surface-mediated programming of network topology can be implemented.

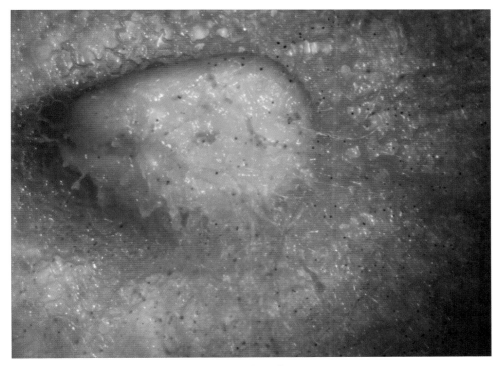

Cooperative organism. ©2014 Ben de Lacy Costello.

It is true that when culturing P. polycephalum, the appearance of fungal hyphae or spores is usually an indication that subculturing is imminently required. However, P. polycephalum will live very happily alongside certain fungi such as the pin moulds pictured. In fact, it is apparent that P. polycephalum utilises the pin mould structure as a scaffold, growing appreciable distances up the stalks of the mould. This same phenomenon is exploited using artificial cilia, to implement pressure and deflection detection.[6] It does not seem to be the case that Physarum utilises the pin mould as a food source; it is possible that it is attracted to populate the stalks by organic compounds secreted by the mould, maybe during sporulation. It was found that Physarum was sensitive to a number of organic compounds which have been associated with fungi.[39]

Demineralisation. ©2014 Ben de Lacy Costello.
Another example of the durability of *P. polycephalum*. Here the organism is happily growing on a cobalt ferricyanide precipitating chemical system. It was able to grow on and around the chemical reaction far beyond the normal expected lifetime of a single inoculated culture, up to 30 days. Presumably it is utilising the biocidal nature of the chemicals against other pathogenic organisms to achieve this longevity. It was also apparent that at the sites of colonisation there was a significant demineralisation of the formed precipitate. This manifested itself first as the formation of Liesegang rings (just visible in the picture) but subsequently as a complete absence of precipitate in certain regions. Certainly, Liesegang structures have never been observed previously on uncolonised precipitates, so Physarum appears to play a role in their formation, presumably the active transport of reactants, products and precursors. Can this phenomenon be usefully harnessed, i.e. for the construction of functional materials?

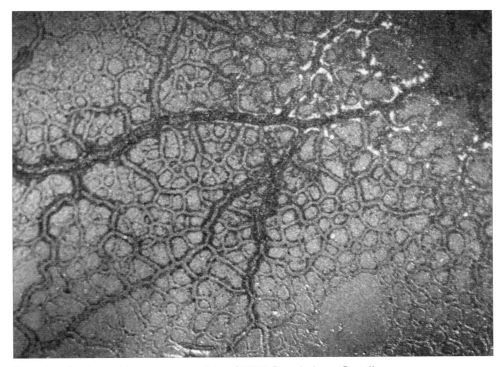

Functionalisation with smart materials. ©2014 Ben de Lacy Costello.
This shows a residual network which was chemically functionalised with active material by allowing the Physarum to grow on a substrate loaded gel (potassium ferrocyanide) followed by direct functionalisation with a reactant (copper chloride). This is in contrast to previous examples where active remote transport of the reactant was utilised as the functionalisation mechanism. The *P. polycephalum* network is functionalised with copper ferrocyanide. Copper ferrocyanide and other Prussian blue analogues which can also be used to functionalise Physarum are well-known functional materials with applications in photocatalysis, electrochromic devices, electrocatalysts, photomagnetic and magneto-optic devices, batteries, ion sensing and photoimage formation. Thus, functionalised networks of *P. polycephalum* have a range of potential uses.

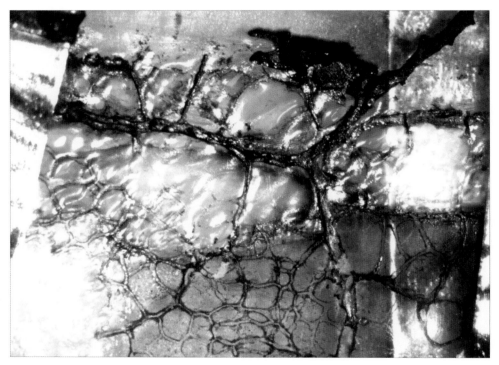

Conducting polymer functionalised. ©2014 Ben de Lacy Costello.
The image shows a Physarum network spanning two aluminium electrodes which has been functionalised using a selective coating mechanism. Local parts of the network can be soaked in an activator (in this case ferric chloride) which promotes polymerisation of the selected monomer (in this case pyrrole) on contact. The image shows a large functionalised tube which bridges the 1 cm gap between the two electrodes. The resistance of the tube was measured to be in the region of 100 kΩ, in line with an intrinsic semiconducting conducting polymer. Using this methodology, it is possible to functionalise Physarum tubes with a wide range of conducting polymers (substituted pyrroles,[40] polythiophenes and derivatives, polyanilines and derivatives). It can be seen from the image that the selective local functionalisation means that certain parts of the network are raw unfunctionalised Physarum which are still alive.

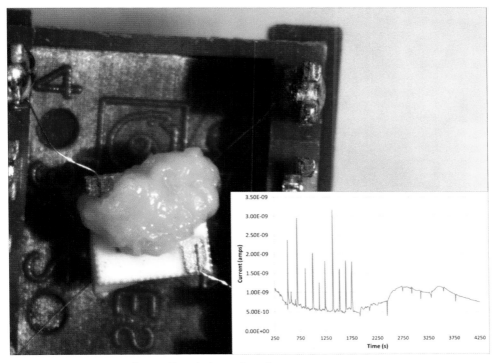

Slime vapour sensor. ©2014 Ben de Lacy Costello.

The plasmodium of the slime mould can be deposited onto a gold interdigitated sensor substrate (2 mm^2). Used in the same way as a conventional vapour sensor,[41] the changes in current in the presence of volatiles can be assessed. The inset shows a current/time plot for 10 exposures to breath, which give an increase in current, and subsequently multiple injections of 50 ppm ethanol vapour, which cause a smaller but sharp decrease in current. There are some baseline fluctuations apparent but the relative time scales of these versus the response time do not interfere unduly with the sensing mechanism. The Physarum was stable in this format for a number of hours and could be rehydrated to increase the longevity of the sensing mechanism.

Chemotaxis towards and away from volatile chemicals. ©2014 Ben de Lacy Costello.

P. polycephalum demonstrates chemotaxis towards or away from a range of volatile substances. In the images shown, Physarum is moving directly towards a source of farnesene and away from geraniol. The Physarum also remains localised within the vicinity of the activator, in this case, for up to 48 hr after inoculation. By undertaking binary assays between a series of volatile chemicals with different functionalities, it was possible to construct an order of preference for Physarum. For compounds causing positive chemotaxis, the order was: farnesene > β-myrcene > tridecane > limonene > p-cymene > 3-octanone > β-pinene > m-cresol > benzyl acetate > cis-3-hexenyl acetate. For inhibitory compounds the order in terms of inhibition strength was: nonanal > benzaldehyde > methyl benzoate > linalool > methyl-p-benzoquinone > eugenol > benzyl alcohol > geraniol > 2-phenylethanol. This work shows that P. polycephalum has a preference for non-polar higher molecular weight compounds such as terpenes.[39]

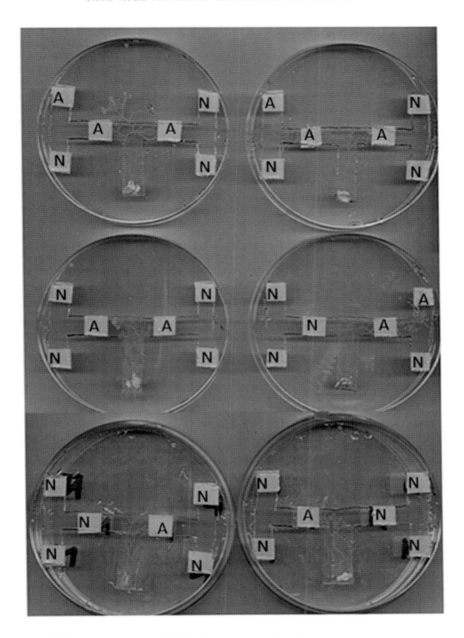

Routing of Physarum signals. ©2014 Ben de Lacy Costello.

If the principle of chemotaxis towards an activator is used, then Physarum "signals" can be routed through various simple junctions. This is based on the preference of Physarum to propagate towards the activator (A) rather than the uncoated filter paper (N). If two activators are present at the same junction, then the signal is split. If activator is present at only one junction, then the signal propagates in that direction.[42]

Chapter 5

Basic features of slime mould motility

Tomohiro Shirakawa

*Department of Computer Science,
School of Electrical and Computer Engineering,
National Defense Academy of Japan, Kanagawa, Japan
sirakawa@nda.ac.jp*

The plasmodium of *Physarum polycephalum* is a unicellular and multi-nuclear giant amoeba that is formed by fusions of myriads of uninucleate microscopic amoebae at a point in the life cycle of the organism. The very large unicellular form of the plasmodium is very uncommon in nature; on the contrary, almost all of the other higher organisms have multi-cellular bodies. Therefore, the plasmodium has an exceptional property: although the plasmodium is a unicellular organism, the size of the amoeba is variable. The smallest plasmodium consists of the fusion of two amoebae, so the smallest size is twice that of a usual amoeba. There is no upper limit to the largest size of the plasmodium, in principle. There is a record of very large plasmodium of more than a few metres. A more interesting point is that despite the variety in the size, the plasmodium can move, feed and form complex structures and adapt itself to the environment in an intelligent manner.

Since the plasmodium has a very large cell body and thus some cellular phenomena can be observed in an enhanced manner, the organism has been used as a model organism in conventional biology to study phenomena such as cell motility and cytoplasmic streaming. Furthermore, nowadays, the plasmodium is also studied as a model of nature-inspired unconventional computing with its adaptive behaviours.[43,44] For example, recent studies revealed that the taxes of the plasmodium enable a spatial optimisation between environmental stimuli and that oscillatory behaviour of the cell body is able to hold a memory of periodic events.[45,46] To investigate the cellular machineries that realise such phenomena is still an important topic. Therefore, in this chapter we illustrate the nature of cellular machineries of the plasmodium, especially with the phenomena that are induced by the large-scale body of the organism.

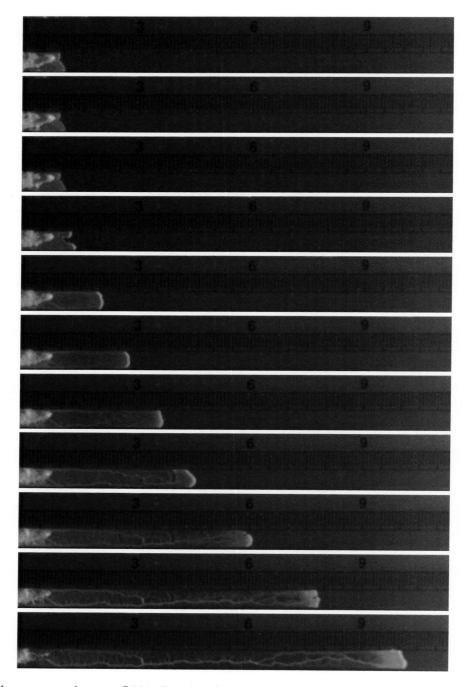

Physarum accelerates. ©2014 Tomohiro Shirakawa.

Time-lapse images of the plasmodium speeding up its propagation proportionally to its body length.[47] The top photograph is the start of the motion and subsequent photographs are made at 100 min intervals. The plasmodium propagates in channels 5 mm wide. Scales are shown in cm. The volume and weight of the plasmodium were constant.

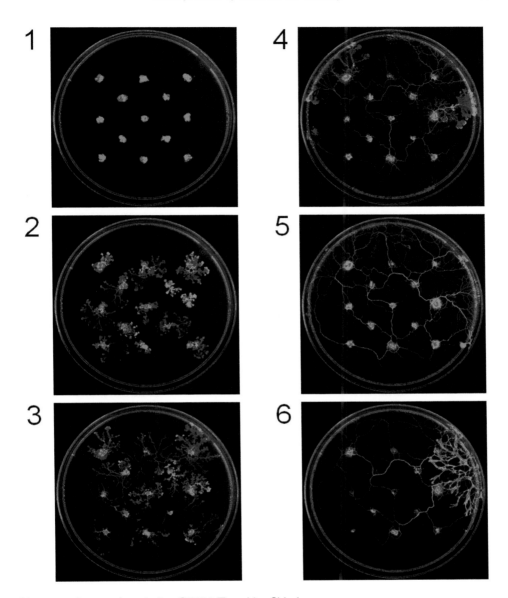

Physarum knows the whole. ©2014 Tomohiro Shirakawa.

The motility of 13 pieces of the plasmodium on a plain agar plate in a 9 cm Petri dish. Image 1 indicates the initial state of the plasmodia and the following images indicate their time development at 180 min intervals. The colours of the plasmodia indicate the thickness and grey, red, green, blue are in the order of increasing thickness. During their growth, the plasmodia come in contact with each other and fuse. The plasmodia do not enter a space which has been explored once. When the whole space is fully explored (image 5), the plasmodium gets out of the agar plate for the first time and intrudes onto the top of the dish, the dry plastic surface of which the plasmodium usually avoids.

Physarum travels: the trajectory of 3-day journey. ©2014 Tomohiro Shirakawa.
The trajectory of the plasmodial motility during 3 days. The height of the figure corresponds to 100 cm. 100 mg of the plasmodium was inoculated on a plain agar surface and photographed at 10-min intervals. Then all the images were binarised and superimposed, with a black background. The plasmodium travelled from the top to the bottom, taking 3 days.[48]

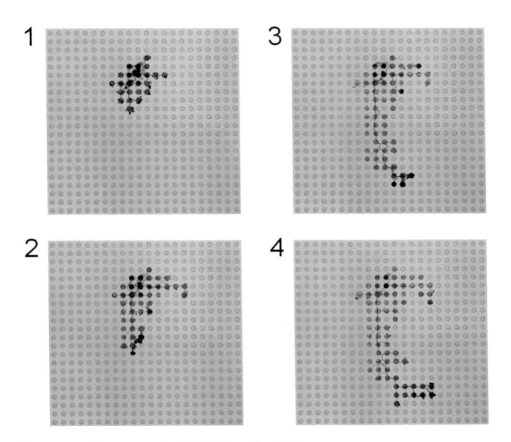

Physarum cellular automata. ©2014 Tomohiro Shirakawa.

The time development of cellular automata made of the plasmodium. The time intervals between the figures are 200 min. 40 mg of the plasmodium was inoculated on a non-uniform substrate that consisted of polycarbonate and plain agar surfaces. Circular fields of agar, 3 mm in diameter, were arranged in a reticular pattern with 5 mm intervals. Since the plasmodium avoids dry surfaces, it tends to pass by the polycarbonate surface quickly and stay longer on the agar surface. As a result, the motility of the plasmodium is limited to four neighbours.[49]

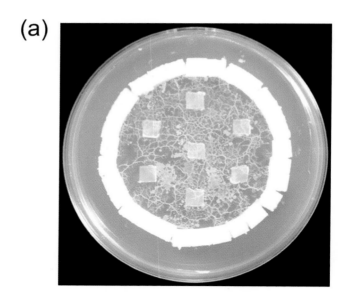

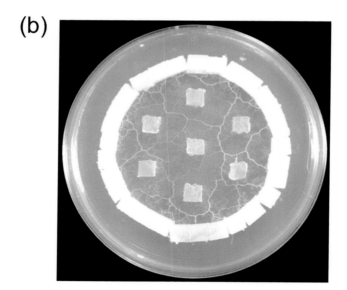

Physarum Voronoi diagram. ©2014 Tomohiro Shirakawa.

Formation of Voronoi diagram by the plasmodium. The experiment was performed on a plain agar plate in a 9 cm Petri dish. In the figure, the white cubes are cubes of nutrient agar, and the transparent cubes are agar cubes with salt. (a) After the plasmodium covered the space, the cubes were placed in the space. The plasmodium moved outside to obtain nutrients, at the same time avoiding the repellents. (b) The tubular network was formed avoiding the repellents as much as possible and, as a result, a Voronoi diagram of a planar point set represented by transparent cubes was approximated.[44]

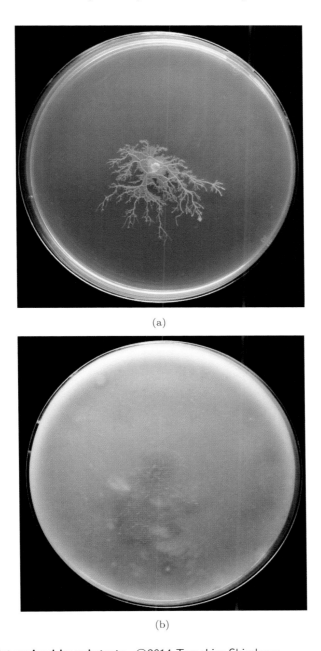

(a)

(b)

Morphology determined by substrate. ©2014 Tomohiro Shirakawa.

Development of the morphology of the plasmodia in different conditions. The experiments were performed on a repellent containing agar (a) and on a nutrient containing agar (b), in a 9 cm Petri dish. The plasmodium formed a branching structure on the repellent agar; on the other hand, it developed a rather disordered structure on the nutrient agar. The dependency of the plasmodial morphology on the substrate has been studied.[50]

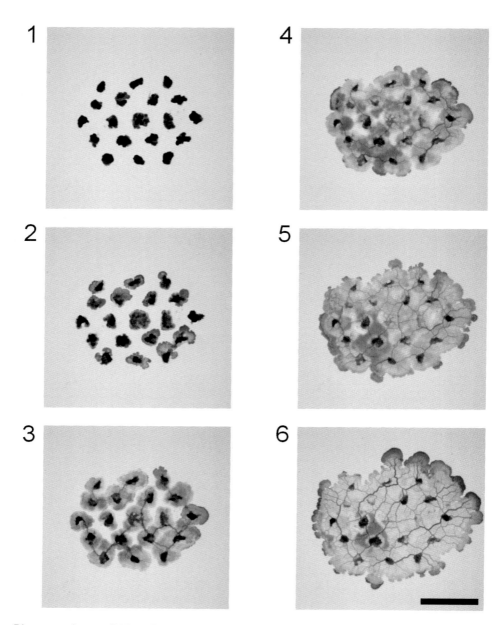

Physarum fuses. ©2014 Tomohiro Shirakawa.

Fusion of multiple plasmodia. The experiment was done on a plain agar plate. Image 1 shows the initial state of 19 plasmodial pieces, and the following figures show the ones at 50, 100, 150, 200 and 230 min later, respectively. The plasmodia spread and come into contact with each other, and then fuse together. After the fusion of all pieces, the plasmodium obtains a unity as a single cell (images 5 and 6). Bar: 1 cm.

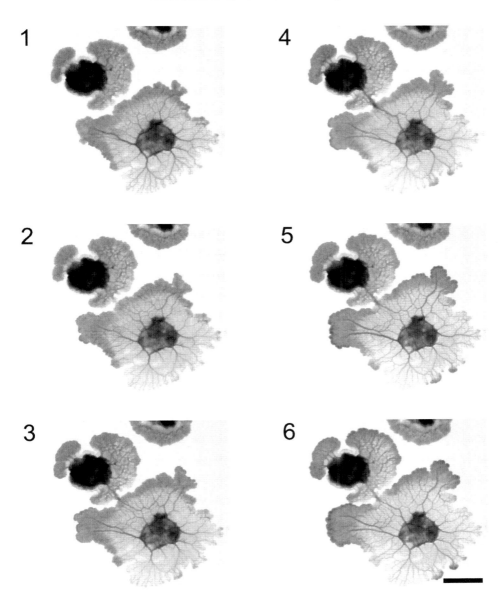

Physarum fusion: the close-up. ©2014 Tomohiro Shirakawa.

Fusion of two plasmodial pieces. The time intervals between the figures are 10 min. As shown in the figures, if the masses of the plasmodia are relatively small and the plasmodia are on nutrient-free or nutrient-repellent substrates, the plasmodia make fusion at one point only and the other parts avoid each other.[43] Bar: 5 mm.

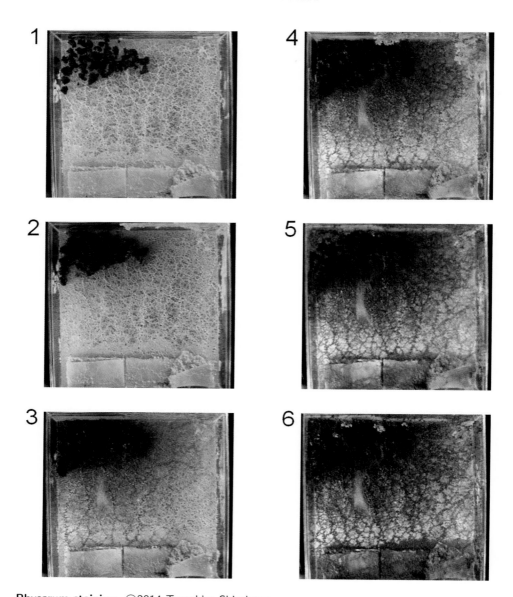

Physarum staining. ©2014 Tomohiro Shirakawa.
Staining of the plasmodium using Indian ink. The time intervals between the figures are 100 min. For the plasmodium fully spreading on a plain agar plate in a plastic box 15 cm by 15 cm, dry oat flakes containing Indian ink were supplied (the black mass in the upper left in image 1). By feeding the oat flakes, the plasmodium incorporated the component of the ink at the same time, and the body of the plasmodium was stained by the ink component. Note that the agar surface is not stained, and this indicates that the propagation of the black colour is not because of diffusion of ink, but because of intracellular transportation inside the plasmodium.

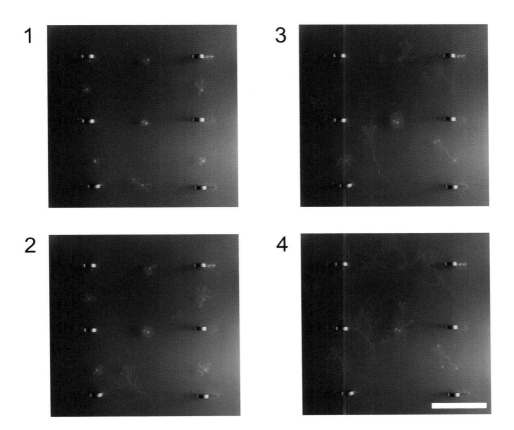

Magnetotaxis of Physarum. ©2014 Tomohiro Shirakawa.
Motility of the plasmodia induced by magnetism. The experiment was performed on a plain agar plate that had six neodymium magnets facing in the same direction on its surface. The time intervals between the figures are 250 min. In the final figure, all of the seven pieces of the plasmodia headed in the same direction, along the direction of the magnetic line, to the south pole.[51] Bar: 3 cm.

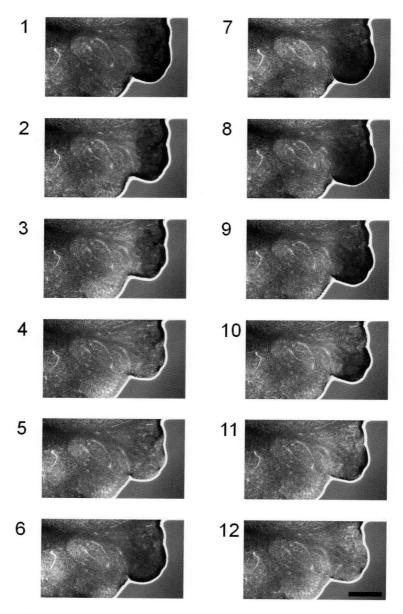

Microscopic images of pseudopod. ©2014 Tomohiro Shirakawa.

Phase-contrast microscopic images of a plasmodial pseudopod. The time intervals between the figures are 10 sec. In images 1–4 and 9–12, the cytoplasm in the pseudopod flows out and the frontal part is shrinking. On the contrary, in images 5–8, the cytoplasm flows into the pseudopod and the frontal part is swelling. Bar: 200 μm.

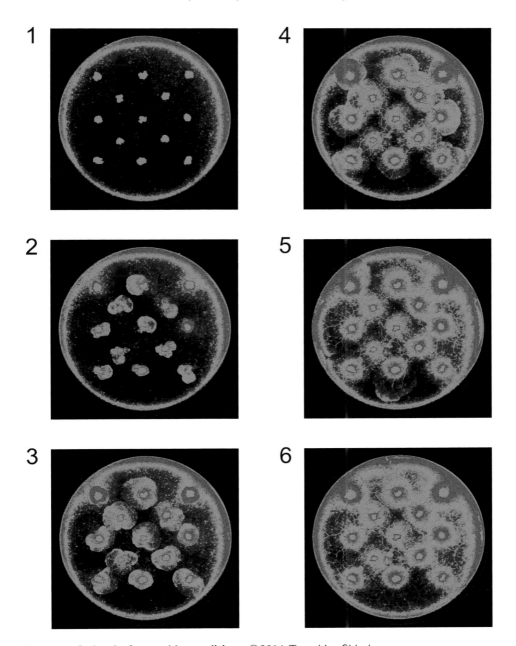

Physarum fusion in favourable condition. ©2014 Tomohiro Shirakawa.

The fusion of the plasmodia in a nutrient-rich condition. The experiment was performed on a nutrient-containing agar in a 9 cm Petri dish. The first figure indicates the initial state of the plasmodia just after the inoculation. The time intervals between the figures are 250 min. The colours of the plasmodia indicate the thickness and grey, red, green, blue are in the order of decreasing thickness, grey is the thickest. In this nutrient-rich favourable condition, the plasmodia spread homogeneously, and the plasmodia are immediately merged just after the contact.[43]

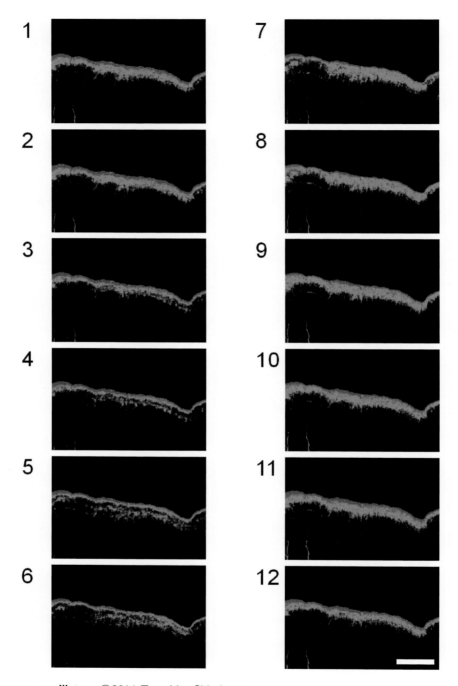

Physarum oscillates. ©2014 Tomohiro Shirakawa.

Pseudo-coloured images of the plasmodial motility and oscillation in thickness on a plain agar plate. The time intervals between the figures are 20 sec; the colours of the plasmodium indicate the thickness and grey, red, green, blue are in the order of decreasing thickness. Bar: 1 cm.

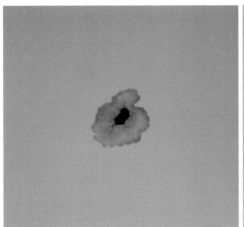

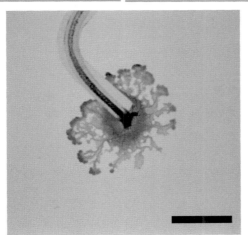

Physarum transplantation. ©2014 Tomohiro Shirakawa.

The morphology of the plasmodium 200 min after the inoculation and transplantation of a tube. In the experiment, 20 mg of the plasmodium was inoculated on a plain agar plate and the tubular structure of the plasmodium was further transplanted: (top left) without transplantation, (top right) 2 cm tube, (bottom) 4 cm tube. The tube of the plasmodium is included in the agar gel tube, and one end of the agar tube is blocked by white petrolatum. As shown in the figures, the tubular structure promoted the development of heterogeneous morphology and cell motility according to its length.[52] Bar: 1 cm.

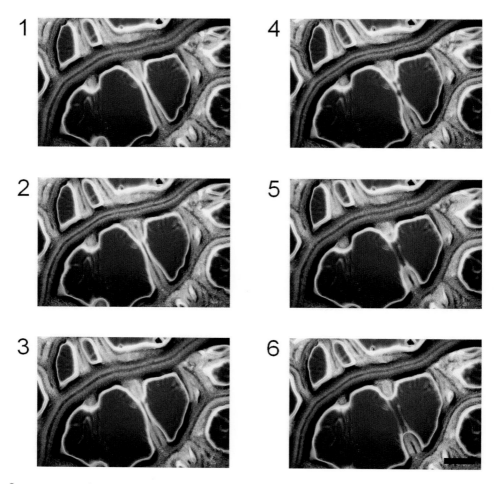

Spontaneous cleavage of tubular structure. ⓒ2014 Tomohiro Shirakawa.

Phase-contrast microscopic images of plasmodial tubular structures. The time intervals between the figures are 30 sec. The tubular structure in the centre was spontaneously cleaved and shrunk. Bar: 200 μm.

Chapter 6

Slime mould grown on polymer layers

Alice Dimonte*, Tatiana Berzina and Victor Erokhin

Institute of Materials for Electronics and Magnetism,
Italian National Research Council, Parma, Italy
alice.dimonte@imem.cnr.it

The term *bioelectronics* can have two meanings: integration of living beings or their elements into electrical circuits artificially or reproduction of some properties of biological objects with electronic components. The applicability of the first approach can be illustrated by the large number of publications related to the realisation of biosensors. These electronic compounds are based on the unique capabilities of biological molecules for specific recognition with very high degree of selectivity and reliability, which is impossible with human-made systems. Usually, biosensors are based on antibodies, enzymes and even cells. In the case of the second approach, the most important example is to realise a system capable of bioinspired or biomimicking information processing. For example, the concept of neural networks is directly related to the realisation of systems capable of learning similarly to living beings. However, the main efforts now in this field were done in the software realisation of such systems. The situation drastically changed in 2008 when a first experimental realisation of the *memristor* was reported[53] The term memristor was introduced by Leon Chua in 1971 by considering the symmetry of electrical circuits.[54] The resistance of this hypothetical element must be a function of the electrical charge that has passed through it: this is very similar to the property of the Hebbian synapse. Regarding the electronic circuits, this can be reformulated in the following way: the conductivity of the connection between nodes with non-linear properties must increase with the duration and frequency of the involvement of this connection in the process of the signal pathway formation. The device realisation of the memristor was a starting point of the increased activity in the field of the neural networks at the hardware level. In particular, several papers reporting neuron[55,56] and synapse[57,58,59,60,61,62] properties mimicking memristors were published. Moreover, an electronic circuit mimicking Pavlov's dog behaviour was realised with memristors.[63] The majority of currently available memristive devices use thin oxide layers as an active medium. The organic memristive device[64] differs from the oxide-based elements. It was developed directly for the mimicking of synapse properties for utilisation as a key element in adaptive networks.[65] Direct demonstration of the possibility of using an organic memristive device as electronic analogue of the synapse was done by the realisation of the electronic circuit reproducing the part of the pond snail nervous system responsible for the learning of the animal during feeding.[64] In this case, compared to the Pavlov's dog mimicking circuit, we have reproduced not only properties but also the architecture of the nervous system. Further studies were based on the unique properties of the organic systems, allowing self-assembling. This property allowed us to realise three-dimensional stochastic networks of memristive devices.[66] Several features of the network were comparable to the learning of the animal brain.[67] In this respect, the ongoing project Physarum chip[20] is a natural combination of the two approaches of bioelectronics mentioned above: we combine biomimicking properties of the organic memristive systems with computational capabilities of the slime mould in resolving optimisation tasks.[32,27,14] As the result, we will have a hybrid living/organic system; the electrical properties of each counterpart will be strongly affected by the other. This chapter will illustrate the progress achieved in some directions related to the project.

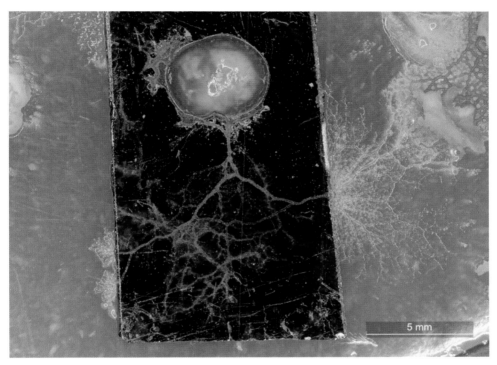

Physarum on polyaniline. ©2014 Alice Dimonte, Tatiana Berzina and Victor Erokhin.

This picture and the one below are related to the same experimental series. A small blob of P. polycephalum, almost 2 μL, has been placed on a silicon wafer slice, previously deposited with commercial polyaniline (PANI) by the Langmuir–Schaefer technique. Samples, with the mould on top, were placed into a Petri dish with 1.5% agar non-nutrient gel. In this way, it was possible to keep the mould in a proper, humid, environment. To induce Physarum to design networks, some oat flakes have been spread on the agar, around the sample. Physarum changes with its own body the conductivity state of PANI layers in different ways, providing negative and positive patterning of the sample. PANI electrochromism arises in a colour outburst related to the addition or extraction of a proton.[68]

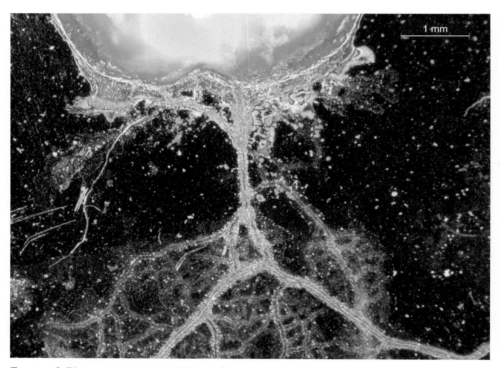

Zoom of Physarum on polyaniline. ©2014 Alice Dimonte, Tatiana Berzina and Victor Erokhin.

This picture features the artistic network designs performed by Physarum as it was branching towards the food through the PANI substrates.

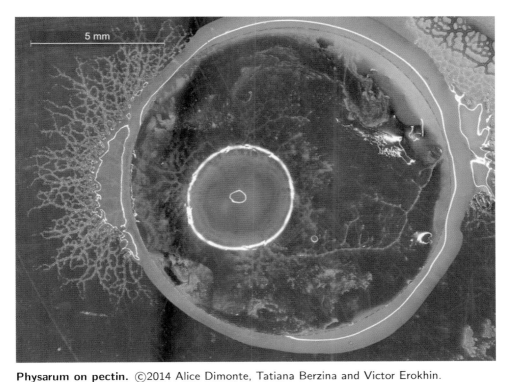

Physarum on pectin. ©2014 Alice Dimonte, Tatiana Berzina and Victor Erokhin.

Physarum has incredible adaptation capability. It redesigns continuously its shape forming optimised networks of protoplasmic tubes. This unusual eye-like shape is the result of Physarum placed on glass with pectin.

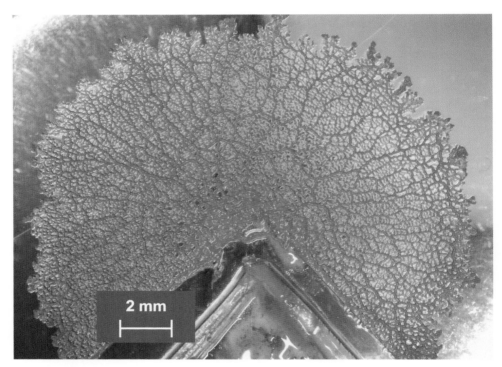

Physarum and xylose lysine deoxycholate agar. ©2014 Alice Dimonte, Tatiana Berzina and Victor Erokhin.

Optical microscope image showing a network of protoplasmic tubes created by Physarum in the presence of a chemical repellent: xylose lysine deoxycholate (XLD) agar. Physarum propagates avoiding the area of the repellent, creating a fan-shaped design.

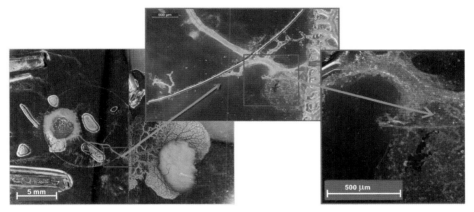

Physarum and polystyrene microparticles. ©2014 Alice Dimonte, Tatiana Berzina and Victor Erokhin.

Optical microscope images of experiments done with Physarum loaded with red polystyrene microparticles (1 μm in diameter). The glass substrate was divided with a stripe of particles, thanks to an *ad hoc*-designed mask and a solution casting deposition technique. A 250 μL Physarum blob was placed on one side of the stripe. The substrate was surrounded on three sides with rods of a chemical repellent (see yellow rods at the top and bottom of the picture on the left) in order to force Physarum's growth only in the direction of the microparticle stripe. The repellent was agar xylose lysine deoxycholate. After several hours, the mould created protoplasmic tubes crossing the stripe and we registered the presence of microparticles at the opposite side of the stripe, engulfing and carrying the particles out of the glass substrate.

Chapter 7

Diversity of slime mould circuits

Martin Grube
Institute of Plant Sciences,
University of Graz, Austria
martin.grube@uni-graz.at

Once fused from single amoebae to form an initial plasmodium, *P. polycephalum* and related species in the order *Physarales* continue growth via synchronous cell division and extension of the megacells by a contractile plasmodial vein network. These display the phenomenon of shuttle streaming, a rhythmic back-and-forth flow of the protoplasm within a tubular system at a period of approximately 1 to 2 min, with the frequency depending on the nutritional benefits. With time, this develops as a microfluidic circuitry that adapts patterns of contraction to size to optimise the transport throughout an organism. This control of fluid also creates dynamic changes in network architecture seen over time in an individual. Food quality is of considerable importance for slime mould growth dynamics.[69]

So far, experiments were conducted almost exclusively with *P. polycephalum*, but further acellular slime moulds also produce macroplasmodia, such as *Badhamia utricularis* and *Fuligo septica*. *B. utricularis* and *P. polycephalum* are quite similar but seem to differ in their growth rate when grown together with oat flakes. The first figure shows one of the competitive experiments, demonstrating the more vigorous growth of *P. polycephalum* under these axenic conditions. The properties of both species have been considered a biological model for concurrent games.[70] While the plasmodia seem to avoid the sharing of the same area in initial phases of these experiments, longer experimental times also show that the two species may explore their habitat independently. In these cases, encounters and crossings of strands belonging to the two species can frequently be observed. There is neither an apparent mutual influence of growth nor a coordination of the shuttle streaming among the species, shown in the second figure. Both species thus develop parallel circuits which do not interfere with each other.

As a third and more distantly related species, *Fuligo septica* not only appears to grow more slowly, but also has less affinity to oat flakes. It is definitely less competitive in confronting assays with *P. polycephalum* and *B. utricularis*. However, the circuits produced by *F. septica* have an interesting property that has not been observed with the two other species. If the agar substrate is superficially carved using a razor blade, *F. septica* tends to follow the carved lines, even at angles of 90°, as shown in the third figure. This property may be used to guide the growth of Fuligo circuits, which may be interesting for unconventional computing approaches.

P. polycephalum also has the property to avoid previously explored areas of the medium, due to repellent activity of the extracellular slime left behind by the moving plasmodium. This self-avoidance is also observed in *Badhamia utricularis* but not in *Fuligo septica*, which rather seems to maintain growth on already secreted slime tracks. The fourth figure shows an experiment with agar plugs already colonised by *F. septica* displaying distinct slime traces. When these plugs are placed on a fresh agar surface in a proper manner, the lack of self-avoidance can be used to reroute the slime mould circuitry in a controlled way. Here a circular slime mould circuit has been constructed. The interesting properties of *F. septica* make this organism interesting for implementing logical gates.[71]

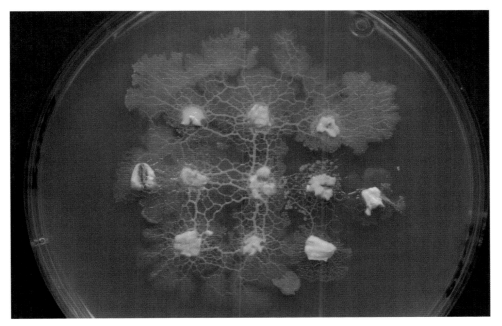

Bioinspired game theory. ©2014 Sigrun Kraker.
P. polycephalum confronting *Badhamia utricularis*. Physarum started from the left, whereas Badhamia started from the right-most oat flake to colonise a grid of nine oat flakes. Here, representative for most experiments, Physarum grows more vigorously and colonised the seven right and upper flakes, while Badhamia colonised only two flakes to the lower left. There is no clear mutual inhibition of the species, but it appears that Physarum enforces its own network to prepare for new paths, rather than attacking the other species.

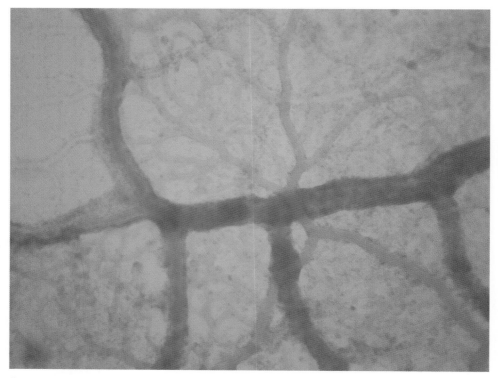

Parallel computing of slime mould species. ©2014 Martin Grube.

Strands of *P. polycephalum* and *Badhamia utricularis* may coexist in an independent manner. Here, the paler coloured strands of Badhamia over-cross the strands of the more intensely yellow Physarum. There is no mutual influence on the shuttle streaming patterns between the species.

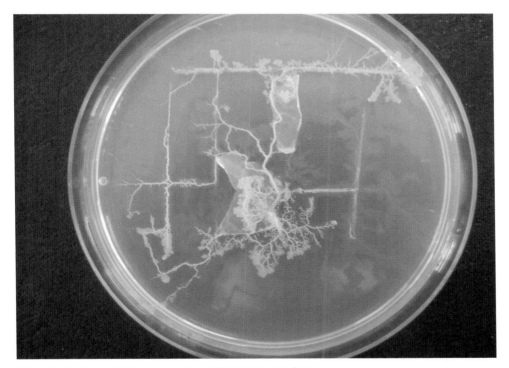

Edge detection with slime moulds. ©2014 Martin Grube.

Fuligo septica, a slime mould in the same family as *P. polycephalum*, forms macroplasmodia which are sensitive to edges in the substrate. Here, the plasmodial strands follow carvings cut with a razor blade in the solid water/agar medium, at angles of even $90°$. The slime of this species contains the tetramic acid derivative fuligorubin A, which is able to chelate metals and is responsible for the massive tolerance of zinc.

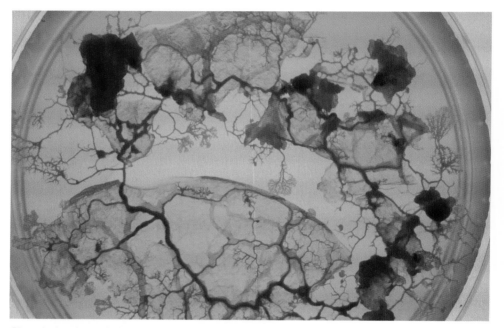

Closed circuitry of slime mould. ©2014 Martin Grube.

Fuligo septica, in contrast to Physarum, does not express self-avoidance. The sensitivity to edges seems to favour growth on its own slime tracks. Together with comparatively slow growth of this species, it is possible to construct circular growth patterns by properly placing already overgrown agar plugs.

Chapter 8

Slime mould fluids and networks from an artist's point of view

Theresa Schubert
Faculty of Media,
Chair of Media Environments,
Bauhaus-University Weimar,
Germany
theresa.schubert-minski@uni-weimar.de

Slime mould is a fascinating creature. It is the largest acellular organism known on this planet and a primordial being that had no need to undergo evolution. In nature, this proves it to be a very successful creature. As a biological curiosity, the species *P. polycephalum* serves as a model for network optimisation and cell motility in scientific experiments. It can be interpreted as an "agent" which distributively solves geometrical problems. The starting points for the experiments displayed here are sophisticated setups or growing environments that allow the organism to propagate and grow, yet under conditions that the artist has predefined, sometimes with interaction between human organisms.

The first images examine the fluidics of *P. polycephalum*, initially in an experiment to investigate a living gate formed from slime mould and the influence of tactile stimulation on the flow direction. Then some screenshots of a video installation investigating the cytoplasmic streaming of the cell are shown. In the experiment on slime mould wires, the Arduino Duemilanova is taken as an example for self-growing wires from living matter. Since their production in 2005, Arduino boards have become one of the most popular and important tools of electronic artists. Hence, the choice of this board is symbolically for its importance. By setting up slime mould to grow the wire connections, it hypothesises about replacing traditional hardware with living organic materials. Hybrid Brain is a living sculpture and a metaphor. It criticises the importance of the human brain and investigates the origins of creativity. The study of Kolmogorov–Uspensky machines refers to a hypothetical machine where its storage unit is based on distributed nodes rather than on linearity as with the Turing machine. Taking this as a reference for an implementation with slime mould, the study asks about alternative possibilities of computing and how slime mould processes information.

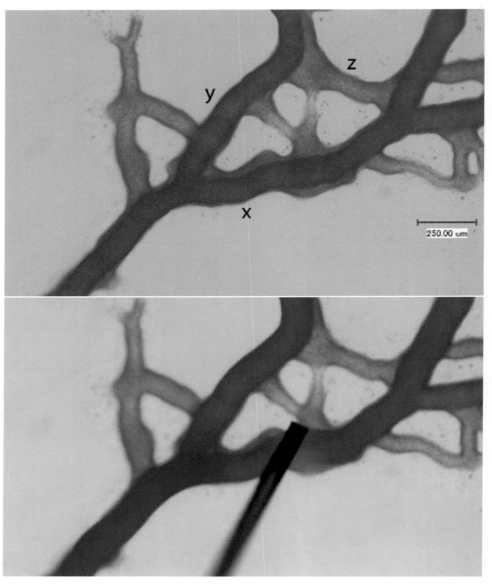

P. polycephalum gate. ©2014 Theresa Schubert.
Implementation of an XOR gate in Physarum. Top: snapshot of the living gate before stimulation, input values are logical TRUTH; input tubes are x and y, output tube is z. Bottom: mechanical stimulation of an input tube.[71]

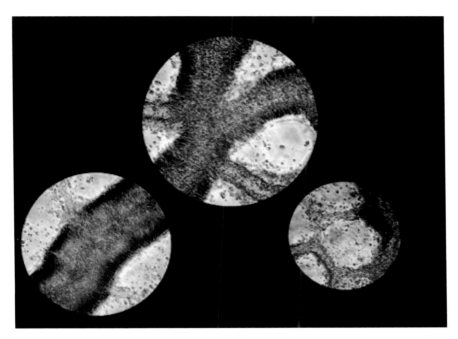

50"flows. ©2014 Theresa Schubert.

Top: 50"flows is a three-channel video projection based on video recordings made at a microscopic level, showing the flow of cytoplasm within the slime mould. The so-called cytoplasmic streaming controls the growth and movement of the "many-headed" protozoa. The resulting network structures are the raw material for an imaginative journey that focuses on the creative potential of simple organisms. Bottom: in this study, the tip of a branch of P. polycephalum was observed under the microscope and the connection between turning of streaming and peristalsis was investigated.[72]

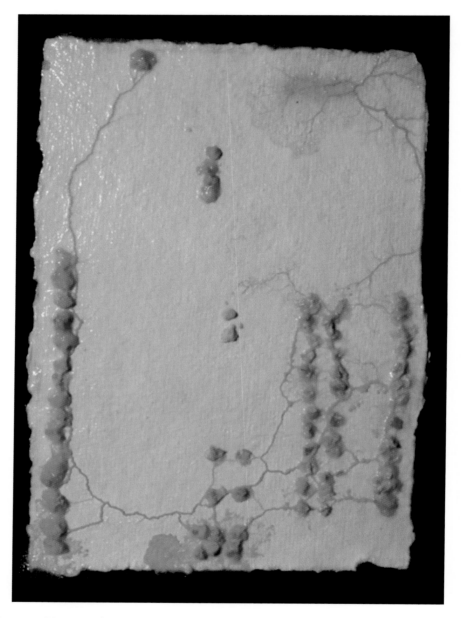

Slime mould wires. ©2014 Theresa Schubert.
On a sheet of hand-made paper (size approx. 10×14 cm^2) oat flakes are placed at the locations of an Arduino Duemilanova board where the solder points from electronic parts are situated. Slime mould is inoculated at the power point (top-right corner). After approx. 20 hr it reaches the microchip (bottom-right area); from there it grows to the middle parts. After 100 hr, it reaches the left-hand output pins and finally after 120 hr, it reaches the USB port at the top-left corner. Pioneering results on the slime mould wires and analyses of their electrical properties have been published.[3]

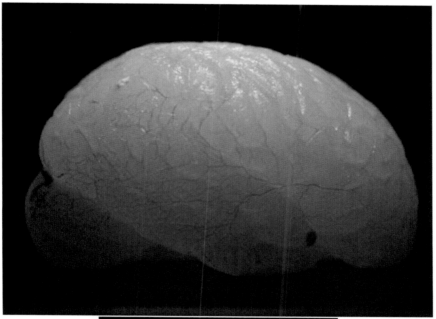

Hybrid Brain. ©2014 Theresa Schubert.

This living sculpture consists of a human brain cast from agar with slime mould growing on top of it. The brain sits in a custom-made vitrine equipped with a water humidifier. Oat flakes are placed at the locations of brain areas responsible for creativity, vision, hearing and smell, which are the zones significant for creating and perceiving an artwork. The slime mould grows its typical vein-like network on top of the brain, forming an external vascularisation. A would-be nervous system made of the slime mould has been presented.[9]

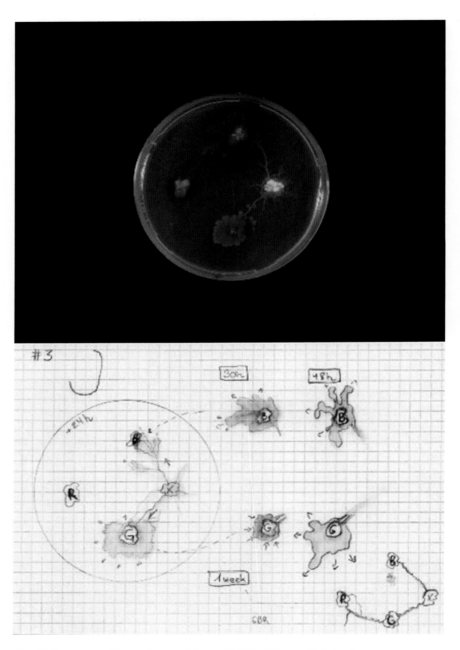

Growing Kolmogorov–Uspensky machines. ©2014 Theresa Schubert.
Slime mould is inoculated in several Petri dishes. The layout of the oat flakes is distributed and colour coded. In contrast to a Turing machine, a Kolmogorov–Uspensky machine has a storage unit which is not an unbounded linear tape but a growing graph.[73] Top: after 36 hr *P. polycephalum* has connected the green and blue flakes. Bottom: drawing with pencil and aquarelle showing the growth over time and the final connected graph.

Bibliography

1. A. Adamatzky, *Bioevaluation of World Transport Networks*. (World Scientific, 2012).
2. A. Adamatzky, Slime mould logical gates: exploring ballistic approach, *arXiv preprint arXiv:1005.2301.* (2010).
3. A. Adamatzky, Physarum wires: Self-growing self-repairing smart wires made from slime mould, *Biomedical Engineering Letters.* **3**(4), 232–241, (2013).
4. A. Adamatzky, Geometry induced delays of slime mould propagation, *Biophysical Reviews and Letters.* **8**(01n02), 89–97, (2013).
5. A. Adamatzky, *Physarum Machines: Computers from Slime Mould*. (World Scientific, 2010).
6. A. Adamatzky, Tactile bristle sensors made with slime mould, *Sensors Journal, IEEE.* **14**, 324–332, (2014).
7. J. G. Whiting, B. P. de Lacy Costello, and A. Adamatzky, Towards slime mould chemical sensor: Mapping chemical inputs onto electrical potential dynamics of *Physarum Polycephalum*, *Sensors and Actuators B: Chemical.* **191**, 844–853, (2014).
8. A. Adamatzky, Slime mould tactile sensor, *Sensors and Actuators B: Chemical.* **188**, 38–44, (2013).
9. A. Adamatzky, A would-be nervous system made from the slime mould, *Artificial Life.* (2014).
10. A. Adamatzky, Developing proximity graphs by *Physarum polycephalum*: Does the plasmodium follow the Toussaint hierarchy? *Parallel Processing Letters.* **19**(01), 105–127, (2009).
11. A. Adamatzky, Physarum machines for space missions, *Acta Futura.* **6**, 53–67, (2013).
12. R. van de Weygaert and W. Schaap. The cosmic web: geometric analysis. In *Data Analysis in Cosmology*, pp. 291–413. Springer, (2009).
13. C. E. Shannon. Presentation of a maze-solving machine. In *8th Conf. of the Josiah Macy Jr. Found.(Cybernetics)*, pp. 173–180, (1951).
14. A. Adamatzky, Slime mold solves maze in one pass, assisted by gradient of chemoattractants, *NanoBioscience, IEEE Transactions on.* **11**(2), 131–134, (2012).
15. A. Adamatzky, Towards slime mould colour sensor: Recognition of colours by *Physarum polycephalum*, *Organic Electronics.* **14**(12), 3355–3361, (2013).
16. A. Adamatzky and R. Mayne, Actin automata: Phenomenology and localizations, *arXiv preprint arXiv:1408.3676.* (2014).
17. NASA, *NASA/JPL-Caltech: Our Milky Way Gets a Makeover.* (NASA, 2008).
18. V. Evangelidis, M.-A. Tsompana, G. Sirakoulis, and A. Adamatzky, Slime mould imitates development of Roman roads in Balkans, *Submitted.* (2014).
19. M.-A. I. Tsompanas, G. C. Sirakoulis, and A. I. Adamatzky, Physarum in silicon: the Greek motorways study, *Natural Computing.* pp. 1–17, (2014).
20. A. Adamatzky, V. Erokhin, M. Grube, T. Schubert, and A. Schumann, Physarum chip project: Growing computers from slime mould., *IJUC.* **8**(4), 319–323, (2012).
21. L. Bull and A. Adamatzky. Evolving gene regulatory networks with mobile DNA

mechanisms. In *Computational Intelligence (UKCI), 2013 13th UK Workshop on*, pp. 1–7. IEEE, (2013).
22. A. Adamatzky, Manipulating substances with *Physarum polycephalum*, *Materials Science and Engineering: C.* **30**(8), 1211–1220, (2010).
23. A. Adamatzky, B. de Lacy Costello, and T. Shirakawa, Universal computation with limited resources: Belousov–Zhabotinsky and Physarum computers, *International Journal of Bifurcation and Chaos.* **18**(08), 2373–2389, (2008).
24. A. Adamatzky and G. J. Martinez, Bio-imitation of Mexican migration routes to the USA with slime mould on 3D terrains, *Journal of Bionic Engineering.* **10**(2), 242–250, (2013).
25. A. Adamatzky, Route 20, autobahn 7 and *Physarum polycephalum*: Approximating longest roads in USA and Germany with slime mould on 3D terrains, *IEEE Trans Systems, Man and Cybernetics.* pp. 126–136, (2014).
26. A. Adamatzky, The world's colonization and trade routes formation as imitated by slime mould, *International Journal of Bifurcation and Chaos.* **22**(08), (2012).
27. A. Adamatzky, Slime mould computes planar shapes, *International Journal of Bio-Inspired Computation.* **4**(3), 149–154, (2012).
28. J. Jones, Characteristics of pattern formation and evolution in approximations of *Physarum* transport networks, *Artificial Life.* **16**(2), 127–153, (2010).
29. J. Jones and A. Adamatzky, Emergence of self-organized amoeboid movement in a multi-agent approximation of *Physarum Polycephalum*, *Bioinspiration and Biomimetics.* **7**(1), 016009, (2012). URL http://stacks.iop.org/1748-3190/7/i=1/a=016009.
30. S. Tsuda, J. Jones, and A. Adamatzky, Towards *Physarum* engines, *Applied Bionics and Biomechanics.* **9**(3), 221–240, (2012).
31. J. Jones and A. Adamatzky, Computation of the travelling salesman problem by a shrinking blob, *Natural Computing.* **13**(1), 1–16, (2014).
32. J. Jones and A. Adamatzky, Approximation of statistical analysis and estimation by morphological adaptation in a model of slime mould, *International Journal of Unconventional Computing.* **in-press**, (2014).
33. J. Jones, A morphological adaptation approach to path planning inspired by slime mould, *International Journal of General Systems.* **in-press**, (2014).
34. E. Lucas, *Récreéations Mathématiques.* vol. 1, (Paris, 1882).
35. R. Mayne, J. Jones, and A. Adamatzky, On the role of the plasmodial cytoskeleton in facilitating intelligent behaviour in slime mould *Physarum polycephalum*, *Communicative and Integrative Biology.* **7**(1), e32097, (2014).
36. R. Mayne and A. Adamatzky, Slime mould foraging behaviour as optically-coupled logical operations, *International Journal of General Systems.* **In press**, (2014).
37. R. Mayne and A. Adamatzky, On the internalisation, intraplasmodial carriage and excretion of metallic nanoparticles in the slime mould, *Physarum polycephalum*, *International Journal of Nanotechnology and Molecular Computation.* **3**, 1–14, (2011).
38. R. Mayne and A. Adamatzky, Towards hybrid nanostructure-slime mould devices, *Nano LIFE.* **4**, 1450007, (2014).
39. B. P. de Lacy Costello and A. I. Adamatzky, Assessing the chemotaxis behavior of *Physarum polycephalum* to a range of simple volatile organic chemicals, *Communicative and Integrative Biology.* **6**(5), (2013).
40. N. Guernion, B. de Lacy Costello, and N. M. Ratcliffe, The synthesis of 3-octadecyl- and 3-docosylpyrrole, their polymerisation and incorporation into novel composite gas sensitive resistors, *Synthetic Metals.* **128**(2), 139–147, (2002).
41. B. de Lacy Costello, R. J. Ewen, N. M. Ratcliffe, and P. Sivanand, Thick film organic

vapour sensors based on binary mixtures of metal oxides, *Sensors and Actuators B: Chemical.* **92**(1), 159–166, (2003).

42. A. I. Adamatzky and B. de Lacy Costello, Routing of *Physarum polycephalum* "signals" using simple chemicals, *Communicative and Integrative Biology.* **7**(1), e28543, (2014).

43. T. Shirakawa, A. Adamatzky, Y.-P. Gunji, and Y. Miyake, On simultaneous construction of Voronoi diagram and Delaunay triangulation by *Physarum polycephalum*, *International Journal of Bifurcation and Chaos.* **19**(09), 3109–3117, (2009).

44. T. Shirakawa and Y.-P. Gunji, Computation of Voronoi diagram and collision-free path using the plasmodium of *Physarum polycephalum.*, *International Journal of Unconventional Computing.* **6**(2), (2010).

45. T. Saigusa, A. Tero, T. Nakagaki, and Y. Kuramoto, Amoebae anticipate periodic events, *Physical Review Letters.* **100**(1), 018101, (2008).

46. T. Shirakawa and Y. Gunji, Emergence of morphological order in the network formation of *Physarum polycephalum*, *Biophysical Chemistry.* **128**(2), 253–260, (2007).

47. T. Shirakawa, Allometric scaling laws in the exploratory behavior of the *Physarum polycephalum*, *International Journal of Artificial Life Research.* **3**(1), 22–33, (2012).

48. M. Nishida, H. Sato, and T. Shirakawa, Analysis for the extensive exploratory behavior of Physarum plasmodium in an extendable space, *Proceeding of the 41st SICE Symposium on Intelligent Systems (Digital).* **C11**(3), (2014).

49. S. Ishiguro, H. Sato, and T. Shirakawa, Analysis of the discretized cell motility of the Physarum plasmodium, *Proceeding of the 41st SICE Symposium on Intelligent Systems (Digital).* **C12**(3), (2014).

50. A. Takamatsu, E. Takaba, and G. Takizawa, Environment-dependent morphology in plasmodium of true slime mold *Physarum polycephalum* and a network growth model., *Journal of Theoretical Biology.* **256**(1), 29 – 44, (2009).

51. T. Shirakawa, R. Konagano, and K. Inoue. Novel taxis of the Physarum plasmodium and a taxis-based simulation of Physarum swarm. In *Soft Computing and Intelligent Systems (SCIS) and 13th International Symposium on Advanced Intelligent Systems (ISIS), 2012 Joint 6th International Conference on*, pp. 296–300. IEEE, (2012).

52. T. Shirakawa, K. Yokoyama, M. Yamachiyo, Y.-P. Gunji, and Y. Miyake, Multi-scaled adaptability in motility and pattern formation of the *Physarum polycephalum*, *International Journal of Bio-Inspired Computation.* **4**(3), 131–138, (2012).

53. D. B. Strukov, G. S. Snider, D. R. Stewart, and R. S. Williams, The missing memristor found, *Nature.* **453**(7191), 80–83, (2008).

54. L. O. Chua, Memristor-the missing circuit element, *Circuit Theory, IEEE Transactions on.* **18**(5), 507–519, (1971).

55. P. Krzysteczko, J. Münchenberger, M. Schäfers, G. Reiss, and A. Thomas, The memristive magnetic tunnel junction as a nanoscopic synapse-neuron system, *Advanced Materials.* **24**(6), 762–766, (2012).

56. D. S. Jeong, I. Kim, M. Ziegler, and H. Kohlstedt, Towards artificial neurons and synapses: a materials point of view, *RSC Advances.* **3**(10), 3169–3183, (2013).

57. S. H. Jo, T. Chang, I. Ebong, B. B. Bhadviya, P. Mazumder, and W. Lu, Nanoscale memristor device as synapse in neuromorphic systems, *Nano Letters.* **10**(4), 1297–1301, (2010).

58. T. Chang, S.-H. Jo, K.-H. Kim, P. Sheridan, S. Gaba, and W. Lu, Synaptic behaviors and modeling of a metal oxide memristive device, *Applied Physics A.* **102**(4), 857–863, (2011).

59. F. Alibart, S. Pleutin, O. Bichler, C. Gamrat, T. Serrano-Gotarredona, B. Linares-Barranco, and D. Vuillaume, A memristive nanoparticle/organic hybrid synapstor for neuroinspired computing, *Advanced Functional Materials.* **22**(3), 609–616, (2012).

60. Z. Q. Wang, H. Y. Xu, X. H. Li, H. Yu, Y. C. Liu, and X. J. Zhu, Synaptic learning and memory functions achieved using oxygen ion migration/diffusion in an amorphous InGaZnO memristor, *Advanced Functional Materials.* **22**(13), 2759–2765, (2012).
61. G. Wendin, D. Vuillaume, M. Calame, S. Yitzchaik, C. Gamrat, G. Cuniberti, V. Beiu, et al., SYMONE project: Synaptic molecular networks for bio-inspired information processing., *International Journal of Unconventional Computing.* **8**(4), 325–332, (2012).
62. D. Kuzum, S. Yu, and H. P. Wong, Synaptic electronics: materials, devices and applications, *Nanotechnology.* **24**(38), 382001, (2013).
63. M. Ziegler, R. Soni, T. Patelczyk, M. Ignatov, T. Bartsch, P. Meuffels, and H. Kohlstedt, An electronic version of Pavlov's dog, *Advanced Functional Materials.* **22**(13), 2744–2749, (2012).
64. V. Erokhin and M. Fontana, Thin film electrochemical memristive systems for bio-inspired computation, *Journal of Computational and Theoretical Nanoscience.* **8**(3), 313–330, (2011).
65. V. Erokhin, T. Berzina, and M. Fontana, Polymeric elements for adaptive networks, *Crystallography Reports.* **52**(1), 159–166, (2007).
66. V. Erokhin, T. Berzina, K. Gorshkov, P. Camorani, A. Pucci, L. Ricci, G. Ruggeri, R. Sigala, and A. Schüz, Stochastic hybrid 3D matrix: learning and adaptation of electrical properties, *Journal of Materials Chemistry.* **22**(43), 22881–22887, (2012).
67. V. Erokhin, On the learning of stochastic networks of organic memristive devices., *International Journal of Unconventional Computing.* **9**(3-4), 303–310, (2013).
68. A. Dimonte, T. Berzina, A. Cifarelli, V. Chiesi, F. Albertini, and V. Erokhin, Conductivity patterning with *Physarum polycephalum*: natural growth and deflecting, *Phys. Status Solidi C* **1–5** (2014)/DOI 10.1002/pssc.2014.
69. T. Latty and M. Beekman, Food quality affects search strategy in the acellular slime mould, *Physarum polycephalum, Behavioral Ecology.* **20**(6), 1160–1167, (2009).
70. A. Schumann, K. Pancerz, A. Adamatzky, and M. Grube. Bio-inspired game theory: The case of *Physarum polycephalum*. In *8th International Conference on Bio-inspired Information and Communications Technologies*, (2014).
71. A. Adamatzky and T. Schubert, Slime mold microfluidic logical gates, *Materials Today.* **17**(2), 86–91, (2014).
72. T. Schubert, *Bio:logic Speculations.* (Exhibition at Basics Festival Salzburg, 2012). URL http://vimeo.com/channels/artscience/37942296.
73. A. Adamatzky, Physarum machine: implementation of a Kolmogorov-Uspensky machine on a biological substrate, *Parallel Processing Letters.* **17**(04), 455–467, (2007).

Index

α-shape, 34
β-myrcene, 72
β-pinene, 72
cis-3-hexenyl acetate, 72
XOR gate, 106
2-phenylethanol, 72
3-octanone, 72
4',6-diamidino-2-phenylindole, 49

acceleration, 76
actin, 48–51
   network, 49, 50
activator, 73
active
   material, 63
   zone, 16
ACTN05(C4), 49, 51
adaptive networks, 92
anti-actin antibody, 49
Asia Highway, the, 29
attractant, 12, 13, 15, 28, 30, 34, 52

bacteria, 59
Badhamia utricularis, 99, 101, 102
Balkans, 4
ballistic gate, 2, 30–32
Belousov–Zhabotinsky medium, 16
benzaldehyde, 72
benzoquinone, 72
benzyl acetate, 72
benzyl alcohol, 72
biosensor, 91
biowires, 1

Boolean logic, 30
Boston, 24
brain, 105
branching structure, 81
bristle, 7, 8

cell motility, 75
cellular automata, 79
chemical stimulus, 2
chemoreceptor, 2
chemotaxis, 72, 73
circuit, 11
cobalt ferricyanide, 68
colonisation, 8, 19, 29
colouring, 15
communication, 3
computational geometry, 1
concurrent games, 99
conductive pathway, 2
conductivity, 48, 53, 54, 91
confocal laser scanning microscope, 49–51, 56
confocal laser scanning microscopy, 48
continent, 29
cosmic web, 1, 19
cranium, 9
cresol, 72
cymene, 72
cytoplasmic streaming, 75, 107
cytoskeletal protein, 51
cytoskeleton, 48, 58

DAPI, 49, 50

decision making, 18
deep-field technique, 58
delay, 21, 22
  element, 1
demineralisation, 68
dendritic tree, 2
dissipative soliton, 16

edge detection, 103
electrical
  charge, 91
  circuit, 91
  current, 6
electrochromism, 93
electrode, 9, 53, 54, 70
electron microscopy, 48
electronic board, 13
endocytotic vesicles, 59
endoplasm, 56
eugenol, 72
excitation wave, 1

Füssen, 25
farnesene, 72
ferric chloride, 70
ferrous sulphate, 65
Flensburg, 25
fluorescent nanostructures, 55, 56
fluorophore, 55, 56
frontal scalp, 10
Fuligo septica, 99, 103, 104
fuligorubin A, 103
functionalisation, 69, 70
Fungi, 67

galaxy, 3, 19
  Cartwheel, 19
gate, 30, 53, 54, 105
  XOR, 106
  ballistic, 2
  Boolean, 2
  NAND, 54

geometrical constrain, 20
geraniol, 72
Germany, 25, 27
glabella, 10
gradient, 28
graph, 29
Greece, 5
growing zone, 10

Hebbian synapse, 91
hexenyl acetate, 72
hull, 34
hybridisation, 57

insulation, 14
iron
  ferricyanide, 63
  II/III oxide, 56, 57
  precipitate, 65

Kolmogorov–Uspensky machine, 105, 110

Langmuir–Schaefer technique, 93
latex nanospheres, 56
LED, 11, 12, 52–54
Liesegang ring, 68
light microscopy, 48
limonene, 72
linalool, 72
living sculpture, 109
localisation, 30
logical gate, 1, 48, 53, 54, 100

m-cresol, 72
macroplasmodia, 103
magnetite, 56, 57
maze, 1, 28
  solving, 47
mechanoreceptor, 2
membrane, 59
memristor, 91

metallic particles, 57
methyl benzoate, 72
methyl-p-benzoquinone, 72
Mexico, 23
microinjection, 55
migration, 23
mineralised circuitry, 56
mitochondria, 59
motorway, 5
mountain, 26, 27
multi-agent model, 35
   network formation, 38
   network growth, 39
   pattern formation, 37
   robotics, 36
      amoeboid movement, 40
      collective transport, 42
      guided amoeboid movement, 41
   unconventional computing, 36
      centroid computation, 44
      concave hull, 45
      path planning, 46
      travelling salesman problem, 43
myrcene, 72

NAND, 54
nanoparticles, 48, 55, 57
nanostructures, 56
nervous system, 9, 92
network, 3
   formation, 63
neural
   network, 91
   pathway, 9
neuron, 91
Newport, 24
nonanal, 72
nucleus, 49, 51
nutrient substrate, 19–21

octamethylcyclotetrasiloxane, 14
octanone, 72
ocular pathway, 9
optical stimulus, 2
oscillation, 7, 9
oxide, 91

p-cymene, 72
PANI, 93, 94
parietal bone, 10
path, 18
patterning, 63
pectin, 95
perylene, 56
phenylethanol, 72
photoreceptor, 2
pin mould, 67
pinene, 72
plasmodial tube, 48, 55, 56, 58, 59
polyaniline, 70, 93
polymethylmethacrylate, 66
polystyrene microparticles, 97
polythiophenes, 70
potassium
   ferricyanide, 64, 65
   ferrocyanide, 69
precipitation, 63
proton, 93
protoplasmic
   network, 15
protoplasmic tube, 3, 8–11, 16, 19, 28, 29, 34, 95–97
proximity graph, 1
pyrrole, 70

receptor, 47
repellent, 12, 15, 17, 52
resistance, 70, 91
reticular pattern, 79
Roman
   Empire, 4
   road, 4

Route 20, 24
routing, 12, 73

salt, 12, 17
selective coating, 70
self-assembly, 92
self-avoidance, 100, 104
self-growing, 105
self-repairing, 11
self-routing, 11
semiconductor, 63, 64
sensorial innervation, 10
sensors, 1
shuttle streaming, 99, 102
signal
   routing, 73
   splitting, 73
silicon, 14
   wafer, 93
Silk Road, the, 29
skull, 10
smart material, 69
soliton, 16
space, 3
stochastic networks, 92
synapse, 91

tactile
   sensor, 7
   stimulation, 105
   stimulus, 2

terpene, 72
tetramic acid, 103
transport, 15
   network, 1
   pathways, 1
transportation, 3
tree, 10
tridecane, 72
tubulin, 48, 50, 51
   antibody, 51
   network, 50
Turing machine, 105

USA, 23, 24, 26

vapour sensor, 71
volatile chemicals, 72
Voronoi diagram, 33, 80

wave, 1
   fragment, 17
   front, 16
wire, 2, 11, 12, 105
   insulation, 14
   routing, 12

xylose lysine deoxycholate agar, 96

YL1/2, 51

Zn, 103